# Synthesis Lectures on Computer Vision

**Series Editors**

Gerard Medioni, University of Southern California, Los Angeles, USA

Sven Dickinson, Department of Computer Science, University of Toronto, Toronto, Canada

This series publishes on topics pertaining to computer vision and pattern recognition. The scope follows the purview of premier computer science conferences, and includes the science of scene reconstruction, event detection, video tracking, object recognition, 3D pose estimation, learning, indexing, motion estimation, and image restoration. As a scientific discipline, computer vision is concerned with the theory behind artificial systems that extract information from images. The image data can take many forms, such as video sequences, views from multiple cameras, or multi-dimensional data from a medical scanner. As a technological discipline, computer vision seeks to apply its theories and models for the construction of computer vision systems, such as those in self-driving cars/navigation systems, medical image analysis, and industrial robots.

Katsushi Ikeuchi · Naoki Wake ·
Jun Takamatsu · Kazuhiro Sasabuchi

# Learning-from-Observation 2.0

## Automatic Acquisition of Robot Behavior from Human Demonstration

 Springer

Katsushi Ikeuchi
Applied Robotics Research
Microsoft
Redmond, WA, USA

Naoki Wake
Applied Robotics Research
Microsoft
Redmond, WA, USA

Jun Takamatsu
Applied Robotics Research
Microsoft
Redmond, WA, USA

Kazuhiro Sasabuchi
Applied Robotics Research
Microsoft
Redmond, WA, USA

ISSN 2153-1056        ISSN 2153-1064  (electronic)
Synthesis Lectures on Computer Vision
ISBN 978-3-032-03444-1        ISBN 978-3-032-03445-8   (eBook)
https://doi.org/10.1007/978-3-032-03445-8

This Springer imprint is published by the registered company Springer Nature Switzerland AG
The registered company address is: Gewerbestrasse 11, 6330 Cham, Switzerland

If disposing of this product, please recycle the paper.

# Prologue

## Learning-from-Observation

Learning-from-observation (LfO) is a system designed to mimic human behavior through observation. Unlike methods such as Imitation Learning and Learning from Demonstration [1–4], which directly replicate human joint trajectories, LfO adopts a different, indirect approach. It segments the trajectories into meaningful parts, comprehends the underlying intent of each segment, and subsequently represents this understanding in *task models*, Minsky's frame-like representations [5]. These representations are then mapped to robotic skills, which are pre-trained in advance, for execution. This indirect imitation enables the system to discern which parts of the trajectory are crucial, allowing it to replicate only those essential segments or apply past experiences to new situations.

LfO also distinguishes itself from other methodologies by relying exclusively on visual information for imitation. While force sensations are commonly taught in robot learning—such as in teleoperation—LfO deliberately omits this aspect. According to Piaget's theory of cognitive development [6], infants undergo a developmental stage in which they mimic their parents' behaviors. In this learning process, infants and their parents share a visual space, enabling the infants to acquire actions through observation. However, due to differences in physical characteristics, the force sensations experienced by parents are not directly shared with their infants, requiring the infants to develop force sensations suited to their own bodies.

During this period, infants refine their actions through repeated performance of visually acquired behaviors, likely because force sensation feedback must be optimized for their own bodies. Similarly, in LfO, force sensations are used exclusively in local reinforcement learning to train skill agents for each robot hardware, effectively addressing hardware discrepancies. As a result, LfO operates through a system where, after deriving a general policy from visual information, skill agents—optimized for their specific hardware—execute the actions.

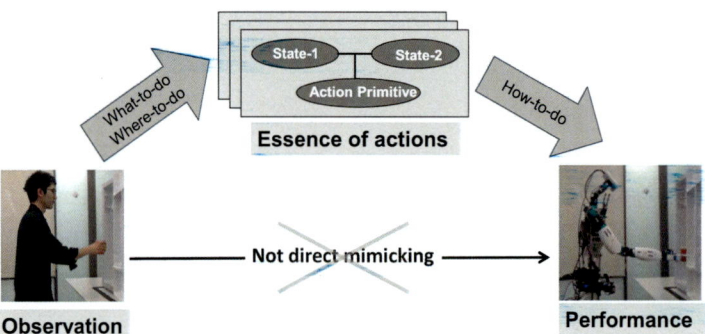

**Fig. p.1**  Learning-from-observation

The term "Learning-from-Observation" itself dates back to the 1990s with the Ikeuchi/ Reddy system [7, 8]. LfO research commenced in the 1990s at CMU with the concept of endowing robots with this ability, analogous to how infants gradually acquire complex behaviors by imitating their mothers. Since then, systems with similar names, such as Learning by Demonstration and Programming by Demonstration, have been developed.

This book defines Learning-from-Observation as an indirect mimicking system where an intermediate framework, referred to as task model, is pre-constructed based on robotics theories. These task models play a central role in both observation and execution; human behavior is observed through the task models, and robot actions are generated based on them (See Fig. p.1).

This book is titled Learning-from-Observation 2.0. LfO has been a subject of research since the 1990s. Significant advancements in machine learning, including large language models, vision-langauge models, and reinforcement learning, have led to remarkable progress in both the encoder, which converts inputs into task models, and the decoder, which executes them. These advancements have resulted in performance that markedly differs from classical LfO systems. Nevertheless, the system retains the name LfO due to its structure, which utilizes an indirect imitation based on the traditional task models, characterized by the distinct separation of encoding and decoding processes.

## Overview of Part I

This book is divided into two main parts. Part I details the functions of each component of LfO 2.0. Part II discusses the evolution of LfO research and the fundamental concepts underlying it. These parts are nearly independent, so you may start reading from either one. Alternatively, you may choose to begin with the prologue and epilogue, followed by the Part II and then Part I.

Chapter 1 of Part I explores the encoder in LfO 2.0, which has undergone significant improvements compared to its predecessor. In the original LfO, the encoder was developed in an ad hoc manner using classical computer vision techniques. However, with advancements in machine learning—particularly in vision-language models—its performance has dramatically improved in LfO 2.0.

The encoder consists of two key components: the task recognizer and the affordance analyzer. Section 1.1 focuses on the task recognizer, detailing how the encoder leverages vision-language models and other advanced technologies to decompose videos of human actions into sequences of task models—such as grasp task model or pick task model—representing what needs to be done. Each task model typically corresponds to a single verb.

A crucial aspect of this system is that the vision-language model is aware of the predefined set of task models (or verbs) available. As a result, it explains video content in terms of these accessible task models. This process of determining "what-to-do" is referred to as task recognition.

The affordance analyzer, described in Sect. 1.2 gathers the parameters essential for task execution and stores them into task models. Initially, it performs grounding to identify the tasks being performed in each segment of the video. Subsequently, it extracts task specific parameters from the corresponding video segments, referred to as skill parameters. For instance, in a grasping task, skill parameters might include the direction from which to grasp the object, while in a picking task, they might specify the direction in which to lift the object. These parameters indicate the position and direction of operations on the target object during task execution and can be considered as the object words corresponding to specific verbs.

The skill parameters, which convey the "where-to-do" information, vary depending on the task. For each task, they are predefined within the task model using corresponding slots, similar to Minsky's frame [5]. This predefined task model is called an *abstract task model*.

The task recognizer selects the appropriate abstract task model from a library of abstract task models. The affordance analyzer then completes the selected model as an "instantiated task model" by acquiring skill parameters. This process of selection and parameter assignment is called *instantiation*.

In programming terms, the task model functions as an operator, with skill parameters acting as the operands corresponding to each operator. The task recognizer selects the appropriate operator, while the affordance analyzer substitutes the operand values. From an object-oriented perspective on task operations, the encoder selects an object, and the affordance analyzer instantiates it. This instantiation triggers daemons within the object's slots, which retrieve the necessary parameters from the provided video segments, thereby completing the object during affordance analysis.

Chapter 2 discusses the decoder, which plays a crucial role in the robot's task execution. The decoder activates a sequence of skill agents based on instantiated task models and their corresponding skill parameters, ensuring consistency across tasks. Each task model has a one-to-one correspondence with a skill agent, which is responsible for executing the required actions on the robot.

Skill agents are pre-trained through reinforcement learning to develop action-specific policies. Consequently, the program units containing these learned policies are referred to as skill agents. They utilize skill parameters, along with visual and force feedback from the environment, to ensure precise execution tailored to each robot's hardware. Once activated by the decoder, skill agents initiate actions using values from human demonstrations as initial parameters and dynamically adjust execution based on feedback—such as force feedback—to achieve the desired behaviors. All available skill agents are stored in a skill library, which is structured to comprehensively cover the target domain without overlap.

Human actions can be categorized into two types: grasping actions, which occur before making contact with an object, and manipulation actions, which take place after contact. As a result, two distinct skill libraries are established—one for grasping actions (Chap. 3) and another for manipulation actions (Chap. 4). These libraries are designed in advance to prevent skill overlap while ensuring comprehensive task coverage through mathematically backed selection processes. Specifically, the grasp skill library is constructed based on Yoshikawa's closure theory (Chap. 3), whereas the manipulation skill library is formulated using the Kuhn-Tucker theory (Chap. 4).

## Overview of Part II

While the Part I discusses the current state of LfO 2.0, Part II conducts a retrospective review of the research trends from the early days.

Chapter 5 in Part II, entitled "Big Bang of LfO," explores the first Learning-from-Observation (LfO) system created by Ikeuchi and Reddy at CMU in 1989 [7]. See also the acompaning YouTube video, *Learning-from-observation: Big bang*.[1] This pioneering system involved a robot replicating human actions to construct identical structures using two blocks. Essentially, it served as an online iteration of the "copy demon" developed by Prof. Mavin Minsky and his students, including Profs. Berthod. K. P. Horn, Patrick H. Winston and Thomas O. Binford, in the 1980s at MIT's AI lab. This system was only capable of handling two blocks, but it introduced the concept of associating state transitions with corresponding manipulation actions—an approach that later became the foundation of LfO's task model design. This chapter further elucidates the parallel structure between this concept and Marr's object recognition framework.

Chapter 6 delves into the concept of LfO in the realm of polyhedral operations. The system introduced in the previous chapter presented the concept of a task model, associated with state transitions, but only dealt with limited transition relations between two cubes.

---

[1] https://youtu.be/UkvDUmRHg4c.

Ikeuchi and Suehiro [9, 10] defined the state of a manipulated object as the range of possible movements due to contact with the environment and defined manipulation actions (i.e., tasks) as causing transitions in this range of movement. This range of movement can be represented as the solution space of a set of simultaneous linear inequalities defined by the surface orientations of the contact points. By examining the rank of the simultaneous linear inequalities using Kuhn-Tucker theory, this solution space can be classified into ten patterns, which consequently defines ten patterns of contact states. Since transitions in contact states occur between these ten states, the upper bound is 100. By sequentially investigating which of these 100 transitions actually occur, they demonstrated that, for polyhedral assembly, only 13 transitions—thus 13 tasks—need to be considered. They prepared 13 abstract task models and constructed LfO in the polyhedral world accordingly.

Chapter 7 discusses LfO in the context of manipulating flexible objects, with a focus on knot tying. Beyond rigid bodies, robots often handle a variety of non-rigid objects. Examples of flexible object manipulation include knot tying and folding clothes.

As an attempt to apply a similar framework in the world of flexible objects, Takamatsu et al. adopted a data format known as P-data, developed in the field of genetic analysis within string theory, to represent the state of ropes [11]. It is known that P-data and rope projections have a one-to-one correspondence, enabling the derivation of unique P-data from projections and the reconstruction of the original projections from P-data. Additionally, transitions in P-data correspond to possible rope manipulations known as Reidemeister moves. By utilizing P-data transitions as task models for rope manipulation, akin to face contact state transitions, they demonstrated the applicability of the LfO framework to flexible objects.

Chapter 8 explores LfO in the context of non-contact tasks, with a particular focus on dance. In polyhedra and ropes, the approach relied on inferring actions based on contact transitions between objects, emphasizing the imitation of outcomes. However, in certain cases, accurately mimicking the movement itself is more important. Dance and martial arts—such as karate and judo—are prime examples.

In these scenarios, exact replication of every movement is impossible due to physical differences between the teacher and the student. Instead, only specific parts of the movements—or the features represented by those parts—are imitated. In other words, within a sequence of continuous movements, only the essential actions that align with the overall purpose, such performing dance or martial arts, are expressed.

In dance, only postures that viewers perceive as identical are imitated. The dance community has developed and used a notation system called Labanotation to describe sequences of these postures [12]. Each Labanotation represents a state, while transitions between them are defined as tasks. Ikeuchi described the upper body using Labanotation [13], while Nakaoka et al. proposed a method for describing the lower body [14, 15]. By integrating these approaches, they developed an LfO system for learning the Aizu Bandaisan dance.

## About Epilogue

In the epilogue, we conclude this book with a discussion on the design philosophy underlying LfO and its distinctions from other methods, approached from an advanced perspective. First, it is argued that LfO has been designed along Piaget's theory of cognitive development, thereby utilizing only visual information and restricting the use of force sensing to local reinforcement learning. Subsequently, the discussion explores why the method of direct imitation, as represented by imitation learning, was not adopted, and what advantages this approach yielded. Finally, the epilogue examines the relevance of Reddy's 90% AI, emphasizing that the design of embodied AI should maintain the fundamental stance of supporting humans, augmenting—rather than replacing—human cognitive and physical abilities.

**Competing Interests** The work presented herein was primarily conducted by the authors in collaboration with their students and colleagues affiliated with Carnegie Mellon University, the University of Tokyo, and Microsoft. The views and opinions expressed are solely those of the authors and do not necessarily reflect those of the institutions. The latest software demonstrated in the videos has been released by Microsoft as open-source under the MIT license.

# Contents

# Part I

# Learning-from-Observation 2.0

This section elaborates on a vision-language model-powered encoder designed to translate human demonstrations into task models. We refer to this computation as an encoder because it converts human actions into a sequence of symbolic representations—i.e., task models.

Within the framework of Learning-from-Observation (LfO), human demonstrations, when observed visually, are interpreted with the granularity of task models, defined in a top-down manner. These task models outline high-level actions and are decoupled from the sequences of motor commands specific to individual hardware. Task models solely encapsulate the "what-to-do" along with the corresponding "where-to-do" parameters, forming representations akin to Minsky's frame [5]. The robot actions necessary to execute these task models—i.e., the "how-to-do"—are provided by a subsequent decoder system equipped with skill libraries.

In this context, the "what-to-do" describes abstract actions that correspond to single verbs.[1] This structure is well suited for general vision-language models (VLM), which function as encoders that describe input scenes using a standardized set of verbs. The VLM-empowered encoder consists of two sequentially connected modules: a task recognizer and an affordance analyzer [16].

The first module, the task recognizer, processes demonstration videos and generates a sequence of task models representing the actions to be executed by a robot. Specifically, this process begins with VLM (VLM) (i.e., GPT-4V) analyzing the video to convert environmental details and actions into text, followed by large language model (LLM) (i.e., GPT-4) generating the corresponding sequence of task models.

---

[1] More precise definitions are given in Chaps. 3 and 4.

© The Author(s), under exclusive license to Springer Nature Switzerland AG 2026     3
K. Ikeuchi et al., *Learning-from-Observation 2.0*, Synthesis Lectures on Computer Vision,
https://doi.org/10.1007/978-3-032-03445-8_1

The second module, the affordance analyzer, determines the "where-to-do" aspect by identifying the timing, locations, and methods of tasks observed in human demonstration videos. It extracts the spatial information necessary for effective robot execution.

Task models can be categorized into two types: abstract task models and instantiated task models, both of which share a structure similar to Minsky's frames. An abstract task model serves as a structural framework with its slots remaining unfilled. First, the encoder identifies an appropriate abstract task model from a set of possible models that correspond to the input action. Subsequently, the affordance analyzer derives the necessary parameters based on the attributes of the slots within the abstract task models, completing its formulation. Once the slots are populated with specific values, the model is referred to as an instantiated task model.

The affordance analyzer reanalyzes specific sections of the video, focusing on the time frames identified by the task model. Through spatiotemporal grounding, it collects affordance data—such as grasp type, way points, and body posture—and combines this data with each task as skill parameters, completing the instantiated task model. This approach can be understood through the lens of sentence analysis, where the affordance analyzer recognizes objective words corresponding to each verb.

It is assumed that input videos for this system are designed for a teaching framework in which the robot replicates human actions. Consequently, there is no self-occlusion or intentional concealment within the demonstrations, ensuring clear and effective task recognition.

## 1.1    Task Recognizer

The task recognizer is further decomposed into three sub-modules: (1) video analyzer, (2) scene analyzer, and (3) task-cohesion generator. See Fig. 1.1.

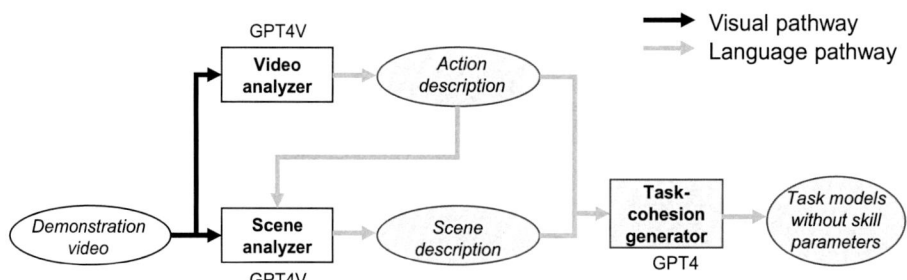

**Fig. 1.1**  Task Recognizer [16]. The input video is captioned using actions that the robot can perform, as identified by the video analyzer. Subsequently, related objects are extracted by the scene analyzer. Finally, the task cohesion generator creates a sequence of task models based on the output of the scene analyzer. The gray components and lines are associated with language/text processing, while the white components and black lines are related to image processing

### 1.1.1   Video Analyzer

When a demo video is input, a VLM recognizes the actions performed by humans in the video and outputs them as text. An example of a VLM is OpenAI's GPT-4V or GPT-4o. To ensure that the subsequent Task Cohesion Generator can effectively process human instruction sentences, the Video Analyzer transcribes the video in a style commonly used for human-to-human communication, essentially functioning as video captioning using a predefined pool of verbs.

Given the token limitations and latency of the VLM, not all frames are analyzed. Instead, video frames are extracted at regular intervals and input into the VLM. By outputting descriptions, human verification and correction become easier.

As an example, we will use a demonstration of a human moving a can of Spam on a table to explain the process. This setup is divided into two parts: an observation station and an execution station,[2] analogous to how a processing system is divided into an encoder and a decoder.

At the observation station, human actions are observed and input into the encoder. Specifically, as shown in Fig. 1.2, a demonstration is conducted in front of the Azure Kinect. AR markers are installed at both stations to establish the rough relationship between their coordinate systems. These markers define a general positional relationship and are used to re-utilize the vector coordinate system of affordance parameters during execution.

It should be noted that during execution, the object's position does not need to be strictly identical at both stations for the execution station to recognize the object's position anew. It is also assumed that the arrangement of obstacles around the target object is almost the same.

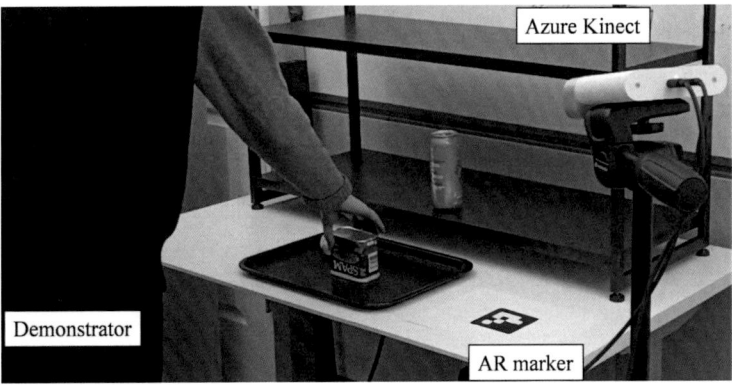

**Fig. 1.2** Observation station

---

[2] Section 2.3 provides a detailed explanation of the execution station, but in summary, the site where the robot executes tasks based on the generated task model is referred to as the execution station.

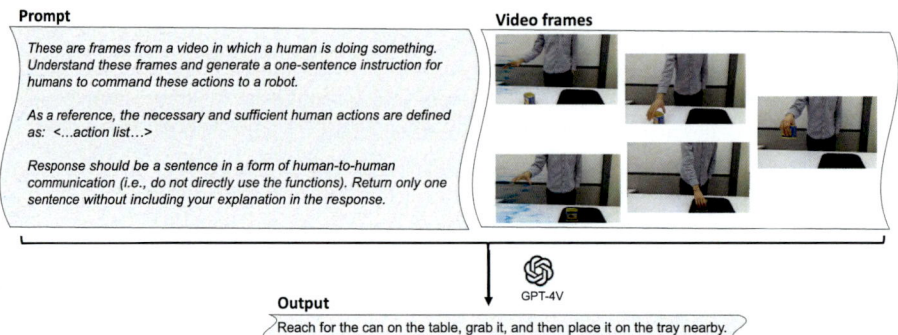

**Fig. 1.3** Video analyzer. The top plane shows the prompt for the GPT-4V, while the bottom plane presents examples of its output. Five frames, including the first and the last frames of the video, are extracted at regular intervals and fed into GPT-4V

As illustrated in Fig. 1.3, the video sequence and prompt are input into the VLM. In front of this observation system, a can of Spam is moved on the tray. The VLM then generates the sentence, "The person reaches for the can on the table, grab it, and then place it on the tray nearby." The role of the video analyzer is to generate this sentence. Notably, unlike general captioning, the description of actions uses only verbs limited by the prompt, rather than common verbs.

### 1.1.2   Scene Analyzer

The scene analyzer converts the work environment into text information based on the text output by the video analyzer and the first frame (or an appropriate frame) of the video data. Specifically, the output includes a list of object names being operated on and the spatial relationships between these objects and related objects.

The object names related to the tasks are provided in an open vocabulary format based on GPT-4V's understanding. These names are then used by the affordance analyzer in the next stage to match with video segments.

Figure 1.4 shows an example of the scene analyzer in action. As illustrated, the scene analyzer identifies objects related to the operation such as "can," "table," and "tray." In this case, since a person moves a Spam can to the tray, the tray is included in the output.

### 1.1.3   Task-Cohesion Generator

The task-cohesion generator uses GPT-4 to generate the necessary sequence of task models from the given action description, environment description, and available abstract task mod-

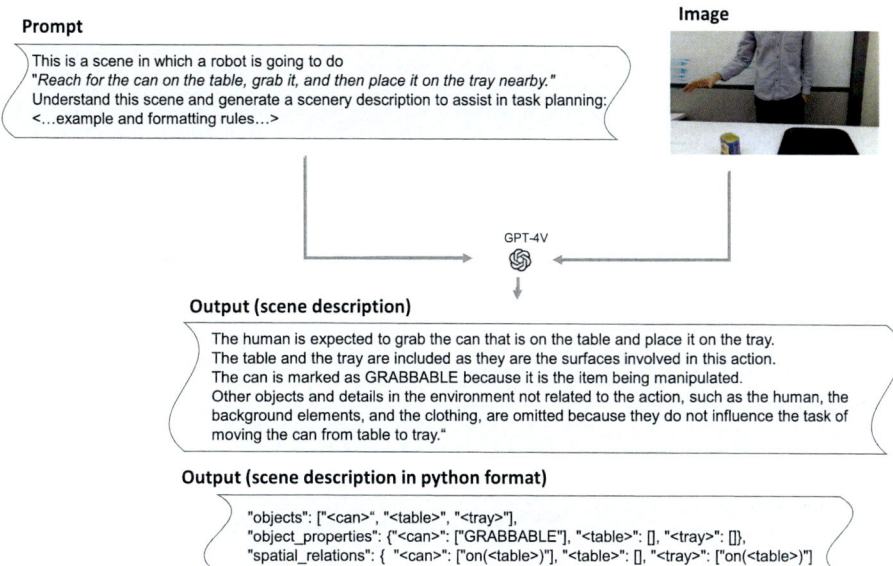

**Fig. 1.4** Scene analyzer. The prompt, along with the output from the video analyzer and the first frame, are provided to GPT-4V. The output list consists of objects related to task cohesion, namely "can," "table," and "tray." Additionally, since the can is being brought, it is assigned the property of "GRABBABLE." Further, the spatial relationships between these objects are also included in the output

els in the library. Here, task cohesion refers to the unity of a sequence of task models. While any sequence of task models exhibits cohesion, this term is specifically used to describe frequently utilized sequences of manipulation task models enclosed by grasp and release task models, with the goal of enhancing the reusability of LfO results.

The abstract task models in the task library, as described in Chaps. 3 and 4, are defined based on Closure theory [17] for the grasping task set, and Kuhn-Tucker theory [18] for the manipulation task set. GPT-4 is provided with prompts detailing the abstract task models available in the library.

To ensure transparency in the GPT-4 model's understanding, the task-cohesion generator is designed to output not only a sequence of task models (task cohesion), but also descriptions of tasks, the estimated environment post-operations, and summaries of the operations. Furthermore, it retains the history of past conversations within GPT-4's token limit. Therefore, users can modify and confirm the output through language feedback based on the results.

Figure 1.5 illustrates an example of the calculation results.

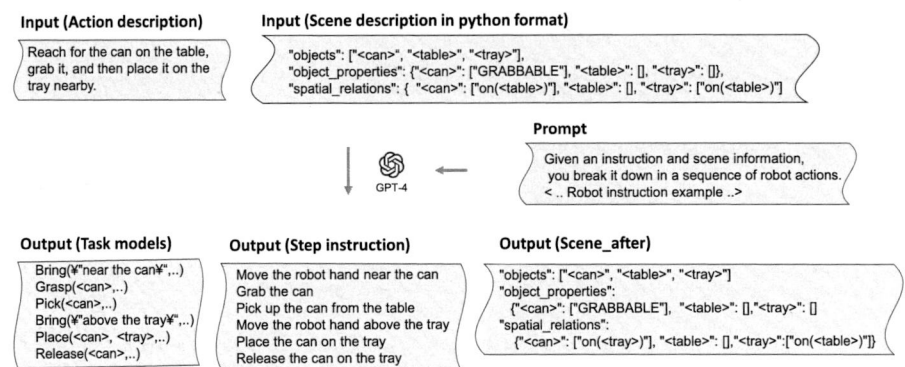

**Input (Action description)**

Reach for the can on the table, grab it, and then place it on the tray nearby.

**Input (Scene description in python format)**

"objects": ["<can>", "<table>", "<tray>"],
"object_properties": {"<can>": ["GRABBABLE"], "<table>": [], "<tray>": []},
"spatial_relations": { "<can>": ["on(<table>)"], "<table>": [], "<tray>": ["on(<table>)"]

**Prompt**

Given an instruction and scene information, you break it down in a sequence of robot actions.
< .. Robot instruction example ..>

GPT-4

**Output (Task models)**

Bring(¥"near the can¥",..)
Grasp(<can>,..)
Pick(<can>,..)
Bring(¥"above the tray¥",..)
Place(<can>, <tray>,..)
Release(<can>,..)

**Output (Step instruction)**

Move the robot hand near the can
Grab the can
Pick up the can from the table
Move the robot hand above the tray
Place the can on the tray
Release the can on the tray

**Output (Scene_after)**

"objects": ["<can>", "<table>", "<tray>"]
"object_properties":
  {"<can>": ["GRABBABLE"], "<table>": [],"<tray>": []
"spatial_relations":
  {"<can>": ["on(<tray>)"], "<table>": [],"<tray>":["on(<table>)"]}

**Fig. 1.5** Task-cohesion generator

## 1.2    Affordance Analyzer for Object Affordance

The affordance analyzer re-examines the given video based on the sequence of abstract task models, task cohesion, to extract the necessary affordance information for the robot's execution (see Fig. 1.6). First, it synchronizes the video with task models using a video-clip selector, an open-vocabulary object detector, and a timing detector.

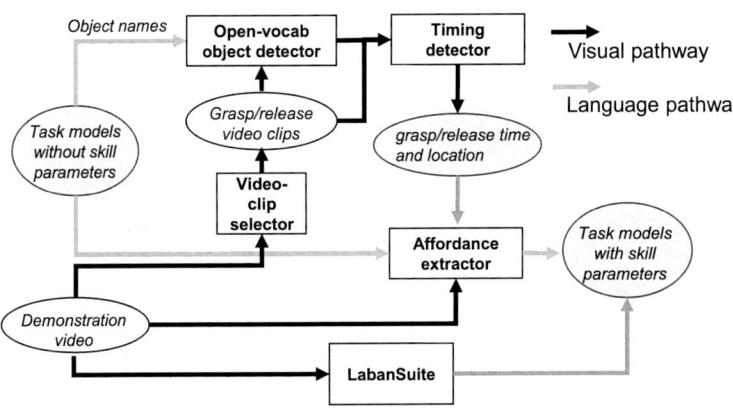

**Fig. 1.6** Affordance analyzer [16]. The affordance analyzer obtains the parameters necessary to complete task models through four modules. It gets the position of the hand and the object position, deriving the relative motion of the hand with respect to the object through Video-clip selector, Open-vocab object detector, and Timing detector. Afforance extractor converts them into skill parameters of the task models by analyzing demonstration video with those information. Concurrently, LabanSuite represents human postures relative to the environment as Labanotation, which is also recorded in the corresponding slots of the task models

Next, the affordance extractor assigns affordance values to each slot within the task models by analyzing the corresponding synchronized video segments. Specifically, this module examines the interaction between the hand and the manipulated object or between the hand and the environment object—such as the approach direction for a grasp task model or the lifting direction for a pick task model—depending on the nature of the task models.

In the current system, it is implicitly assumed that any operation begins with a grasping action, followed by multiple manipulation tasks, and finally concludes with a releasing action, completing the sequence. Thus, the analyzer uses the moments and locations of grasp and release as anchors to synchronize the video with the sequence of task models. It then sequentially analyzes the video segments enclosed between grasp and release tasks, mapping each segment to the corresponding task model. The effectiveness of attention focus in detecting grasp and release has been demonstrated in prior studies [19].

In this book, the affordance information of individual objects recorded in the task model is referred to *skill parameters*. While this information is observed as affordance during the observation phase, it is used as execution information by the skill agent during the execution phase. Therefore, affordance information and skill parameters can be considered identical.

### 1.2.1   Video-Clip Selector

The video-clip selector determines which clips from the segmented video correspond to grasping and releasing. The input video is divided into segments at regular time intervals. The start and end frames of each segment are analyzed using hand detectors and image classifiers and are subsequently classified into the following patterns:

- **Grasp clip**: A segment where nothing is grasped in the initial frame, but something is grasped in the final frame
- **Release clip**: A segment where something is grasped in the initial frame, but nothing is grasped in the final frame
- **Other clip**: segment other than Grasp or Release.

This classification allows the detector to determine which video clips contain moments of grasping and releasing. For this implementation, a YOLO-based hand detector and recognizer [20] is employed.

### 1.2.2   Timing Detector

The timing detector determines the exact timing of grasping and releasing by analyzing the grasp and release video segments. To identify the object intended for grasping, an off-the-shelf Detic open-vocabulary object detector [21] is utilized, leveraging the object names provided by the preceding task-recognizer module. When multiple object candidates are identified, the object closest to the hand within the video segment is designated as the grasped object. The precise moment at which the hand and the object are in closest proximity is identified as the moment of grasping. A similar methodology is employed to ascertain the timing of the releasing action.

### 1.2.3   Affordance Extractor

After aligning the grasp and release timings using the timing detector, temporal correspondence is established to identify the start and end moments of other tasks that intervene between the grasp and release. Our approach assumes stop-and-go demonstrations, where the hand stops momentarily at the boundaries of tasks. For such videos, we segment the video based on the hand speed detected by Kinect. We then perform video recognition for each segmented section and conduct temporal correspondence based on the semantic distance between the recognition results and the verbs representing each task [22].

Following this alignment, affordance information related to the object for each task is extracted. In this example, the tasks include grasp, release, bring, pick, and place. Consequently, affordances are extracted from the video based on the hand movements associated with these tasks.

Figure 1.7a illustrates an example of object affordances. Concerning the grasp task, both the hand utilized and the type of grasp employed are identified as object affordances. Additionally, the direction from which the object is approached is specified. To ensure the reusability of the task model, the approach vector is depicted with respect to the object-centered coordinate system, and the location of the object is not recorded. At runtime, the object name

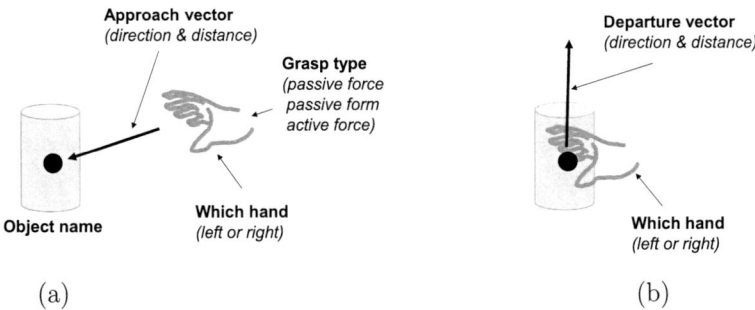

(a)                                                                        (b)

**Fig. 1.7** Examples of object afforance. **a** Object affordance in a grasp task. Regarding positional information, only the approach direction is recorded. The vector is expressed relatively, based on the object-centered coordinate system, to facilitate task-model reuse. It is important to note that the object location is not recorded. **b** Object affordance in a pick task. The departure vector is also expressed relatively with respect to the object-centered coordinate system

in the task model is used to relocate the target object in the environment. From the depth image within the object's bounding box, the object-centered coordinate system is extracted–just as it was during the affordance extraction. The grasp task is recalculated based on this coordinate system.

Figure 1.7b illustrates the object affordances of the pick task. At the time of pick execution, the hand and the object are already integrated due to the previous grasp task. Therefore, only the direction in which the hand moves is recorded as an object affordance. Additionally, this departure vector is represented in the object-centered coordinate system. Consequently, during execution, the object location obtained during the previous pick task is used. This inheritance relationship is maintained by the decoder described in the next section.

## 1.3    LabanSuite for Environment Affordance

In addition to the object affordance mentioned in the previous section, the human posture at the start of each task is obtained as an environmental affordance. Humans are believed to unconsciously adopt an overall posture optimized for the flow of tasks. For example, when grabbing an item from a closed cabinet, one reaches for the cabinet's door handle, extends the arm to open the door, and adopts a posture optimal for the task flow. As shown in Fig. 1.8, two postures can be considered for opening the door. Both configurations are equivalent for the task of opening the door. However, in the flow of sequential tasks, opening the door with the right hand prepares for the next task, which involves reaching inside the cabinet with the left hand to grasp an object. In this case, the configuration shown in (b) is more advantageous for executing the sequence. Therefore, we utilize the sequence of human postures derived from implicit human intentions to resolve the redundancy of humanoid robots and optimize the overall movements during task execution.

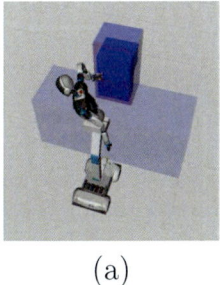
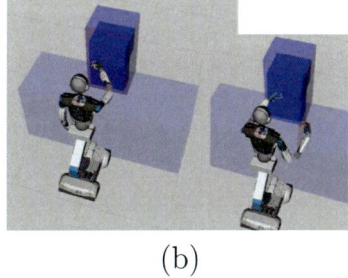

(a)                                                            (b)

**Fig. 1.8** Examples of environment affordance. Two possible postures for opening a refrigerator door. **a** A posture arbitrarily obtained by IK. **b** A better posture when considering the next task (indicated by the human posture)

An indirect strategy is employed to map human postures to robot postures through Labanotation. When considering the adoption of a demonstrated posture by a robot, it is crucial to recognize that, due to differences in joint structures, humans and robots cannot assume identical postures. Rather than striving for exact replication, we endeavor to emulate postures that appear "essentially the same." To obtain these similar-looking postures from human movements, we utilize LabanSuite [13], which generates Labanotation [12] from the movements. Subsequently, these "essentially the same" postures are mapped to robot postures.

Lananotation is a system long used within the dance community to record dance movements, analogous to the use of music scores in documenting musical compositions[3]. Just as the same musical composition can be reproduced from the same music score, the same dance can be recreated from the same Labanotation score. Here, "essentially the same" means that although individual musicians or dancers may exhibit slight variations in their performances, the audience perceives them as essentially identical. For example, in Tchaikovsky's ballet Swan Lake, the Swan Dance, while varying slightly in the details of movements among different dancers due to differences in height and weight, is recognized by the audience as the same dance.

LabanSuite is a system that describes approximate human postures using Labanotation [12]. The posture determined by LabanSuite is utilized as an initial condition for calculating IK. In the case of a general six-degree-of-freedom manipulator, the arm configuration is uniquely determined with respect to the hand's position and orientation. However, for humanoids, the redundancy allows multiple arm configurations from different whole-body postures. We leverage this environmental affordance to reduce such redundancy.

LabanSuite is designed based on a similar sampling method as the notation rules of Labanotation. Labanotation involves sampling in both temporal and spatial dimensions. Although it may appear that humans recognize movements continuously, our research indicates that, at least in the context of dance recognition of Japanese folk dances, humans tend to be aware of and recognize only postures at moments when the movement subtly pauses, called "Tome" [23]. Discussions with ballet dancers have further revealed that when they record ballet dances in Labanotation, they sample postures during these "Tome" moments. Consequently, for the temporal sampling of movements using LabanSuite, the postures at points where human movement slightly pauses are utilized as sample data.

Labanotation, and LabanSuite as well, divides spatial samples into eight horizontal directions and five vertical directions. Although this may seem coarse, according to Miller's law, the human ability to pinpoint a direction within a given interval—the human spatial sampling function—is limited to seven plus or minus two [24]. This sampling function limit applies to each dimension, meaning seven plus or minus two for both the horizontal and vertical directions. The dance community likely chose five directions for the vertical (upward, diagonally upward, horizontal, diagonally downward, and downward) and eight directions for the horizontal (forward, diagonally right forward, right, diagonally right backward, back-

---

[3] Further discussions on Labanotation and LabanSuite can be found in Chap. 8.

ward, diagonally left backward, left, and diagonally left forward) considering symmetry. Consequently, LabanSuite also adopts this number of divisions for spatial sampling.

The description of human posture in the task model as Labanotation scores, namely environmental affordance, is conducted at the beginning of each task using LabanSuite output. LabanSuite converts the given series of human 3D skeleton data into Labanotation dance scores according to the labanotation rules. The series of 3D skeleton data can be obtained directly from Kinect, or in the case of 2D video sequences, using OpenPose and lifting programs. In this system, we utilize skeleton data sequences obtained from Kinect. For these continuous skeletal movements, the moments where movement subtly pauses are taken as temporal sample points, and each joint direction is classified into one of the eight or five directions to obtain the Labanotation score.

Figure 1.9 illustrates the analysis results of a cooking scene using LabanSuite. The moments when a human subtly pauses their movement, such as picking up a pot lid (0.55 s) and placing it on the table (1.80 s), are extracted as sample timings, and the arm posture at these moments is described using Labanotation in Fig. 1.9c. Figure 1.10 is an example of the interpretation of the obtained Labanotation. The extraction of Labanotation from human movements during demonstrations is conducted concurrently with affordance extraction and

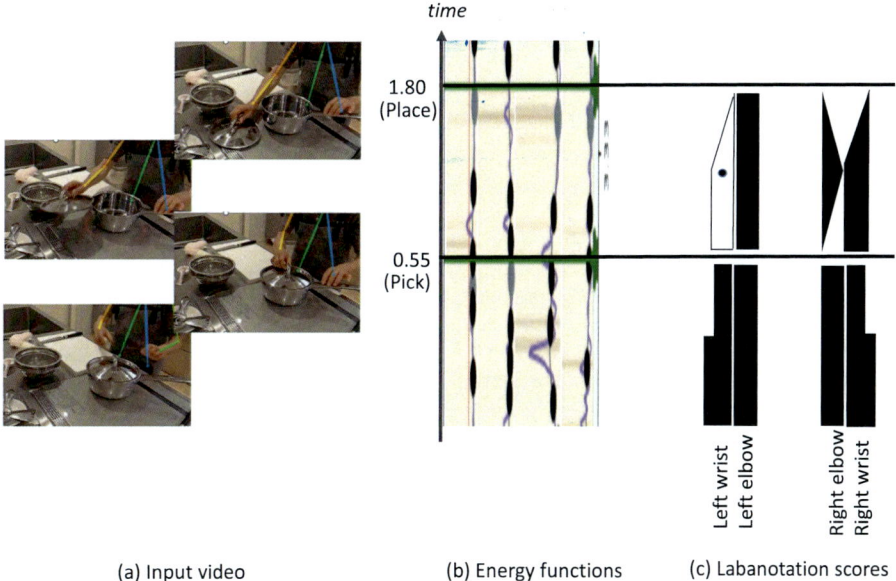

(a) Input video          (b) Energy functions          (c) Labanotation scores

**Fig. 1.9** Analysis of cooking movements using LabanSuite. **a** Input video. **b** The analysis of Laban-Suite. LabanSuite calculates the velocity and acceleration of various parts of the body, extracts extreme points, and determine the sampling timing of movements from their comparison. See [13] for details. Here, 0.55 and 1.80 s have been determined as the timings. These coincide with the Pick and Place timing in the video. **c** Obtained labanotation scores (arm only). At the sampling timing, the posture is converted into Labanotation symbols according to the rules of spatial sampling in Labanotation

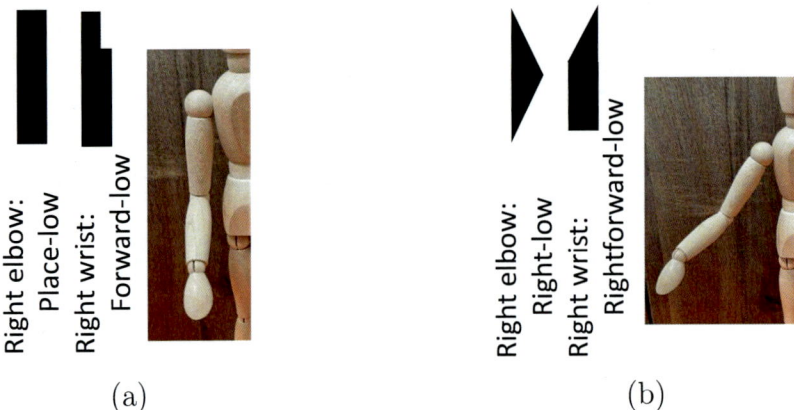

(a)                                                                (b)

**Fig. 1.10** An example of interpretation of Labanotation at the pick and place timing. **a** Pick posture. Regarding the posture at the timing of the pick task, since both symbols are black, they indicate a downward direction in terms of the zenith angle. The black rectangle for the right elbow signifies straight down, while the polygon for the wrist, indicating forward, means it extends slightly downward and forward. **b** Place posture. The black triangle, being black, indicates a downward direction and, due to its triangular shape, signifies to the right. Therefore, the elbow extends downward to the right. On the other hand, the trapezoid indicates a slight rightward direction, meaning the wrist extends slightly to the rigtht and slightly downward

the Labanotation closest to the start time of each task, as output by the affordance analyzer, is selected as the posture for executing each task and recorded in the task model.

## 1.4    Instantiated Task Models

Affordance information is encapsulated as skill parameters within each slot of the abstract task model, completing the instantiated task model. These instantiated task models are then passed to the decoder module for execution.[4] The main body of the decoder is handled by the TSS platform explained later in Sect. 2.2. Since the skill parameters (affordances) are expressed in relative positions with respect to the object's location, the decoder translates these into the world coordinate system by determining the object's location during execution. In addition, the TSS facilitates the adjustment of skill parameters in a series of task models. Subsequently, the decoder invokes pre-trained skill agents, which have been refined through reinforcement learning, to generate the robot's movements. Table 1.1 delineates the abstract

---

[4] The use of two vocabularies, "affordance" and "skill parameters," is intended to illustrate the conceptual distinction whereby the encoder interprets human actions as affordances, and the observed affordance information is then utilized as execution parameters by skill agents during execution. However, for all practical purposes, both terms can be regarded synonymous.

**Table 1.1** The task models with their skill parameters related to the Spam relocation demo. For all possible task models, refer next chapter

| Task model | Skill parameter (Affordance info) | Description |
| --- | --- | --- |
| Grasp | Object | The name of the object to be grasped |
| | Grasp type | Grasp type, either *Passive force*, *Passive form*, or *Active force*, based on the closure theory, to be explained in the next section |
| | Hand | Which hand to use, either *left* or *right* |
| | Approach | Directional vector of the hand approaching the object, while avoiding collisions with the environment. Note that this directional vector has a length, which determines the location where this task begins relative to the object location |
| | Labanotation | Human posture at the beginning of this task using Labanotation |
| Release | Object | The name of the object to be released |
| | Grasp type | Grasp type, either *Passive force*, *Passive form*, or *Active force*, based on the closure theory |
| | Hand | Which hand to use, either *left* or *right* |
| | Departure | Directional vector of the hand departing from the object, while avoiding collisions with the environment. This vector also has a length, indicating the distance the hand will be from the object as a result of this task |
| | Labanotation | Human posture at the beginning of this task using Labanotation |
| Pick | Hand | Which hand to use, either *left* or *right* |
| | Departure | Directional vector of the hand departing from the environment while this task. Since the object is already grasp, it is not specified. This departure vector also has a length, indicating the distance how far the object will be lifted |
| | Labanotation | Human posture at the beginning of this task using Labanotation |
| Bring | Hand | Which hand to use, either *left* or *right* |
| | Goal | Goal position of this task. Assuming the collision avoidance problem has been solved by the demonstrator, the hand with the object grasped moves linearly from the starting location to this goal location |
| | Labanotation | Human posture at the beginning of this task using Labanotation |

(continued)

**Table 1.1** (continued)

| Task model | Skill parameter (Affordance info) | Description |
|---|---|---|
| Place | Hand | Which hand to use, either *left* or *right* |
| | Approach | Directional vector of the hand approaching the environment while this task. Since the object is already grasp, it is not specified. This vector also has a length, indicating the location of starting this task |
| | Labanotation | Human posture at the beginning of this task using Labanotation |

task models and their respective skill parameters related to the Spam-Can demo utilized as an example.

## 1.5    Related Work

### 1.5.1    GPT-Based Encoder

Robot Task recognizer planning based on natural language instructions has been a research topic since the early days of AI [25] and efforts in this direction continue to this day [22, 26, 27]. The advent of large-scale language models in recent years has rapidly accelerated research into their application for robot task planning [28–40]. An excellent survey in this field is presented in [41].

On the other hand, many of these approaches are limited to tasks such as pick-and-place operations [31, 37, 42, 43], or depend on hardware [32, 36, 38, 44, 45]. Additionally, a significant number of these studies rely on specific datasets [28–30, 34, 35], which pose challenges when attempting to adapt or extend them to other robotic setups, often requiring data re-collection or model re-training.

In contrast, the method discussed in this chapter is characterized by its ability to adapt to various operational environments using off-the-shelf general-purpose large language models. Without the need for extensive data collection or model retraining, robot task planning can be created through few-shot learning. Furthermore, as will be elaborated in the next chapter, this approach leverages a skill library validated for necessity and sufficiency, which significantly reduces the likelihood of failure—a key advantage of this method.

## 1.5.2 Affordance Analysis

"Affordance," a term proposed by Gibson, refers to the "meaning" or "opportunity for action" provided by the environment in relation to human behavior [46]. In this context, it denotes the way in which the presence of an object impacts an operator, prompting specific actions at particular locations. This interaction between the operator's purpose and the object plays a pivotal role in enabling the successful completion of intended tasks.

In the framework of LfO, the application of affordance involves first identifying the demonstrator's purpose and determining which opportunities for action were utilized during the demonstration. Rather than comprehensively observing all movements performed by the demonstrator, the approach focuses on understanding the demonstrator's purpose and recognizing the multiple possibilities afforded by the object at specific times and locations. Attention is selectively directed toward the relevant aspects associated with these possibilities. This process is systematically guided by the task model, which explicitly outlines the purpose and affordances associated with each designated slot within the model.

Recognition utilizing the control of focus-of-attention (FoA) is referred to as "active recognition" in the context of computer vision. It is a methodology that enhances image recognition by leveraging FoA [47]. Active recognition is characterized by its use of higher-order knowledge about what the system needs to recognize, controlling FoA in crowded scenes to efficiently process visual information. Ikeuchi and colleagues proposed a task-oriented cognitive system that systematically modifies the structure of the visual system according to the specifications of each task [48].

Applications of linguistic FoA (Focus of Attention) are predominantly limited to specialized domains, such as visual object search [49]. Concurrently, research has explored modeling visual FoA—commonly referred to as 'saliency'—through rule-based or learning-based methodologies [50–53]. More recently, multimodal learning approaches to image processing have emerged as a significant area of interest within the deep learning community (e.g., [54]).

Linguistic FoA itself represents a form of active recognition that leverages linguistic instructions to discern the purpose of a given task. However, as Tsotsos [47] observes, numerous frameworks in this domain fail to dynamically adjust how, where, and when recognition should occur in alignment with the contextual relevance of task objectives. Instead, these frameworks often statically limit the scope of processing, which stands in stark contrast to the adaptive and dynamic methodologies characteristic of active recognition proposed in this chapter.

# Decoder and Its Operation

The decoder processes the sequence of instantiated task models generated by the encoder, adjusting the parameters between these models. Parameter adjustments include correlating the hand's final position in the preceding task with the initial position in the subsequent task, calculating the object's location in real-time, and converting skill parameters described in relative coordinates to absolute coordinates based on these calculations.

Next, the decoder invokes the skill agents responsible for generating the actions of the robot corresponding to each task model. These agents control the robot's movements. The skill agents are pre-trained during the system design phase, prior to the runtime execution phase, and are housed in a skill library. Understanding the detailed workings of the system requires comprehension of this library's structure. However, for the sake of smooth explanation from the encoder to the decoder within the runtime system flow, this section will first describe the decoder's actions based on the task models, and then provide the detailed design of the skill library in Chaps. 3 and 4.

The decoder employs a role-division algorithm, which determines the whole body posture, including arm's posture for humanoid robots using labanotation while maintaining hand movement generated by each skill agent. Each skill agent utilized by the decoder determines the movement of the robot's hand. For a typical six-degree-of-freedom robot, the arm's posture can be uniquely determined by solving the inverse kinematics (IK) from the hand configuration. However, for the humanoid robots targeted in this book, there is redundancy, allowing for multiple possible arm postures. Therefore, the decoder independently determines the arm's posture based on the labanotation described in the task model and resolves the consistency with the hand movement from the skill agents using the remaining degrees of freedom through a role-division algorithm [55].

© The Author(s), under exclusive license to Springer Nature Switzerland AG 2026  19
K. Ikeuchi et al., *Learning-from-Observation 2.0*, Synthesis Lectures on Computer Vision,
https://doi.org/10.1007/978-3-032-03445-8_2

## 2.1    Role Division Mapping

The role division mapping first converts a given Labanotation in a task model into the robot's arm posture using forward kinematics (FK). At this time, it is known that the number of effective direction combinations for the human upper arm and forearm (the number of named postures) is 40 when limited to operations in front of the body (excluding operations behind the back), using an $8 \times 5$ discretization [56]. Consequently, for each robot, joint angles corresponding to the Labanotation are prepared using FK, based on the robot's specific structure.

The role division mapping employs waist, wrist, and lower body movements to ensure consistency between hand movements and arm configuration. Initially, the arm shape determined from the Labanotation is translated parallel to the hand position provided by the skill agent to establish the shoulder position. Furthermore, the wrist joint angle is calculated to align the hand configuration. Meanwhile, since the current position of the robot's lower body is determined by its relative position to the object, consistency between the shoulder position and the lower body position is ensured using the waist joint. Finally, based on the hand configuration provided by the skill agent, these joint angles are used as an initial solution for Bio-IK to fine-tune the overall joint angles.

## 2.2    TSS Platform

The execution of task models on robots, specifically decoding, is conducted on the  TSS platform [57]. Initially named the Task Sequence Simulator (TSS) due to its development as an environment for reinforcement learning of individual skills, it has since become more pertinent to runtime functionalities. Consequently, to avoid confusion with the full name "Task Sequence Simulator," it is now referred to by the abbreviation "TSS."

### 2.2.1    TSS Runtime Functionalities

Figure 2.1 illustrates the functions of TSS. TSS comprises two main functionalities: runtime and compile time. The runtime functionalities can be categorized into four key elements: LfO decoder for parameter consistency, skill controller for whole-body joint angle generation, orchestrator for visualization, and adaptor for robot control.

**LfO decoder** The decoder adjusts and ensures the consistency of skill parameters between task models. TSS receives a sequence of task models from the encoder and maintains the consistency of skill parameters between them. For instance, it aligns the hand position at the start of one task with the end position of the previous task. Additionally, skill parameters are expressed relative to objects and the environment based on an object-centered coordinate

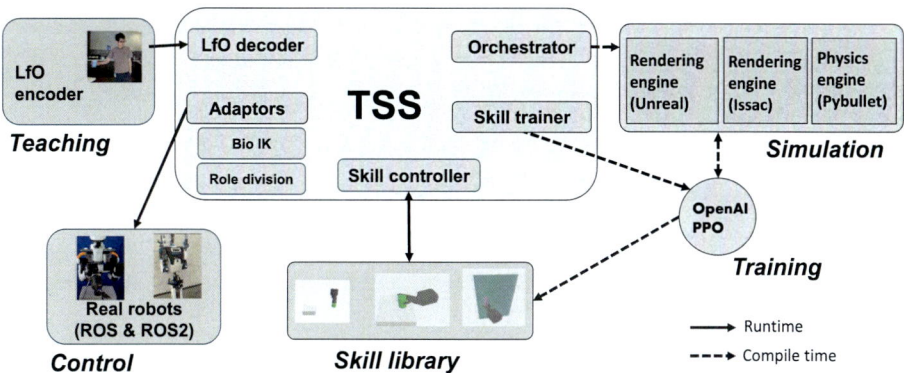

**Fig. 2.1** TSS platform

system. These parameters are then transformed into the world coordinate system of the robot, based on the current positions of target and environmental objects, to enable the robot to execute tasks effectively.

**Skill controller** TSS manages the execution of skill agents by invoking them from the skill library based on the sequence of task models provided by the decoder. Each skill agent functions as an independently operating entity, initializing with the values in the task model and determining the next command value (currently, the next hand configuration) based on learned policies derived from visual and force sensory information from the environment. The controller receives the command value and forwards it to the adapter. When a skill agent sends a termination signal to the controller, the next agent is activated according to the task sequence, repeating the cycle.

**Adaptor** Each robot is connected to TSS through a plug-in socket called an adapter. The inputs to the adapter are the next command value (currently, the hand configuration) and the whole-body configuration represented in Labanotation. Based on these values, the whole-body joint angles are calculated using a role-division and Bio-IK algorithm. The joint angles are then sent to the robot via ROS or ROS2, enabling the robot to operate. Currently, TSS has adapters for humanoid robots such as Kawada-Nextage, THK-Seednoid, and KHI-Nyokkey, as well as mobile robots such as Fetch, Mir and Spot. Additionally, by replacing only these adapters, it is possible to accommodate new humanoids.

## 2.2.2 TSS Compile-Time Functionalities

**Skill trainer** In parallel with these runtime functions, TSS also includes the environment to conduct offline reinforcement learning for each skill. The original name "TSS" was chosen because it was developed to enable the reinforcement learning of a specific skill within a given task sequence. For reinforcement learning environments, PyBullet [58] can be used as physics simulator, while a parallelized version of the Proximal Policy Optimization (PPO) algorithm [59] serves as the learning engine.

**Skill library** The skill library, along with its retrieval and control, is also a feature of TSS. Related to LfO, TSS includes the grasp skill library and the manipulation skill library. In addition to these libraries for object manipulation, there are libraries for generating gestures that correspond to conversation content (gesture library) and for navigation (navigation library). This enables the robot to adapt its roles according to the needs of different scenarios.

**Orchestrator** TSS is equipped with integration capabilities for various simulators. Currently, it can be connected to *Unreal* and *Isaac* for rendering, *Pybullet* for physical simulation. By using the generated whole-body joint angle command values, it is possible to visualize the robot's movements within Unreal and pre-verify its actions.

## 2.3    Execution on Robot Hardware

This section illustrates how the TSS controls actual robots in the execution station. As previously discussed, the LfO system comprises an observation station, which observes demonstrations and encodes them into task models, and an execution station, where the robot performs those tasks by decoding the task models. These stations do not require co-location; they can be situated anywhere, provided they are remotely connected such as by using Azure, which transfers the task models. In this instance, the observation station illustrated in Fig. 1.2 is established in Redmond, while one execution station is positioned in Tokyo, and another in Redmond.

The Nextage humanoid robot, manufactured by Kawada in Japan, is installed at Tokyo execution station (see Fig. 2.2a). The Nextage is equipped with two arms, each possessing six degrees of freedom. Additionally, it has one degree of freedom at the waist. Currently, the lower body has no degrees of freedom. In this experiment, only the right arm is utilized. The right arm is equipped with a six-axis force sensor and a Shadow Hand with 20° of freedom and four fingers. The head is equipped with a ZED sensor, mounted through a two-degree-of-freedom joint at the neck, which is used for object and environment recognition at runtime.

Two robots, Seednoid and Fetch, are used at the Redmond execution station (see Fig. 2.2b and c). The Seednoid, manufactured by THK in Japan, has dual arms, each with seven degrees of freedom. Additionally, it possesses three degree of freedom at the waist and two degrees of freedom in the lower body. In this experiment, only the right arm of the Seednoid was used, which is equipped with a six-degree-of-freedom force sensor and a THK hand with one degree of freedom and three fingers. The head is equipped with a ZED sensor, mounted through a three-degree-of-freedom joint. Meanwhile, Fetch, manufactured

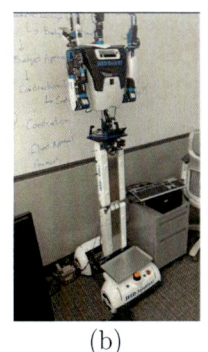
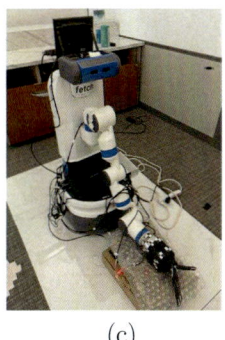

|  (a)  |  (b)  |  (c)  |

**Fig. 2.2** Execution station. **a** Nextage at Tokyo station. **b** Seednoid at Redmond station. **c** Fetch at Redmond station

by Fetch Robotics in the US, has a single arm with seven degrees of freedom, mounted on a two-degree-of-freedom mobile base. The head is equipped with a Prestine sensor, mounted through a two-degree-of-freedom joint.

These robots are connected to the TSS via ROS by the adaptor, and the TSS is further connected to Azure. Each TSS is equipped with a bio-IK and skill library specific to the Nextage with the Shadow hand, a bio-IK and skill library for the Seednoid with THK hand, and a bio-IK and skill library for the Fetch with parallel hand. This configuration allows the same task models to be executed across all the execution stations.

## 2.3.1  Batch Mode (Tray GMR Task-Cohesion)

This subsection elaborates on the decoding process using the task sequence described in the previous chapter–moving a can from the table onto a tray–as an illustrative example.

The subtitle "Batch model" indicates the use of the encoder in batch mode, distinguishing it from the interactive mode, which will be discussed later. In batch mode, the system observes the entire Grasp-Manipulation-Release (GMR) operation from start to finish, generates a sequence of task models with skill parameters, processes them through the decoder, and executes them on the robot. This batch mode is the method described so far in the preceding chapters.

The acronym GMR stands for Grasp-Manipulation-Release, and a GMR task-cohesion refers to a cohesive manipulation task unit bounded by a grasp and a release. While any sequence of tasks can be learned through Learning from Observation (LfO), it is often convenient to define GMR task-cohesion as a modular unit for the purposes of naming and reuse. In this context, the task sequence of moving a can from the table to the tray is designated as the "Tray GMR task-cohesion," which can be reused to transfer an object from one location to another in future scenarios.

Based on the video showing the action of placing a spam can from the table onto the tray, as described previous Sect. 1.1.1, the GPT-based encoder selects the necessary abstract task models from the previously mentioned set of abstract task models in Sect. 1.4 and generates the following sequence of instantiated task models.

```
Bring    {hand}{goal}{labanotation}
Grasp    {object}{hand}{grasp-type}{approach}{labanotation}
Pick     {hand}{departure}{labanotation}
Bring    {hand}{goal}{labanotation}
Place    {hand}{approach}{labanotation}
Release {object}{hand}{grasp-type}{departure}{labanotation}
Bring    {hand}{goal}{labanotation}
```

We will examine how each task model and corresponding skill agent operate along this task sequence.

**Bring task model** The first `Bring` task in the sequence corresponds to the action of moving the hand near the object as a preliminary step before grasping in the Tray GMR task-cohesion. The Bring task model includes the parameters of which hand to move, {hand}, where to move the hand, {goal}, and the human body's posture at the start timing of this task, {labanotation}, as its skill parameters.

```
Bring    {hand}{goal}{labanotation}
```

In correspondence with this `Bring` task model, the skill controller activates the `Bring` skill agent within the library. At the beginning of this activation, the hand is at the end position of the previous task, which, in this case, is the robot's standard position. The objective of the Bring skill agent is to move from the current position to the target position specified in the {goal} slot. To enable this movement, the interval between the final and initial positions are divided, and intermediate target positions are calculated sequentially. The purpose of this segmentation is to use fine intervals to avoid abrupt posture changes caused by the redundancy of inverse kinematics (IK), thereby ensuring that the end-effector maintains as continuous a posture as possible throughout the motion. This position information and {labanotation} are then passed to the role division algorithm, which calculates the full-body posture. The joint angles are subsequently transmitted to the robot's control system via ROS. This loop repeats until the final position is reached. Once the target position is achieved, the `Bring` skill agent terminates and returns control to the skill controller. The controller then activates the agent corresponding to the next task, which, in this case, is the `Grasp` agent.

**Grasp task model** For the `Grasp` task, the task model includes skill parameters such as the object name, the hand to use, the type of grasp, the approach vector, and the posture in Labanotation.

```
Grasp {object}{hand}{grasp-type}{approach}{labanotation}
```

The `Grasp` skill agent responsible for executing this task consists of two distinct modules. The first module, referred to as `Find`, is tasked with determining the approximate position of the object and performing preparatory movements to position the robot body such that the hand can reach the object. In the preceding `Bring` task, the hand was guided to a position near the object in a manner that avoided collisions and enabled grasping. However, at this stage, the robot has not yet recognized the precise location of the target object, and it remains uncertain whether grasping is feasible from the current hand position. Therefore, the `Find` module is employed here to re-identify the object's current location and to formulate a grasping strategy. Subsequently, the second module moves the hand to an approach position relative to the object and executes the actual grasping procedure.

Initially, the "find" module calculates the bounding box of the object in the scene using an off-the-shelf vision system, which is based on the object's name. From the depth image within this bounding box, the distance between the current robot and the object is estimated. This distance estimation allows the robot body to move into an optimal position from which the object can be effectively grasped.

The second module in the `Grasp` skill-agent is responsible for executing the actual grasping operation after the robot has been moved to the designated position by the find module. As detailed in Chap. 3, three distinct pipelines corresponding to the three grasp types—`Passive-force`, `Passive-form`, and `Active-force`—have been prepared. The skill parameter, {Grasp}, within the `Grasp` task model, determines which pipeline should be activated.

Each pipeline comprises two sub-modules: observation and execution. The observation sub-module calculates the contact points of the contact web specific to the grasp type during the grasping process. Here, the contact web represents the distribution of contact points between the fingertips and the object's surface specific to the grasp type. The execution sub-module then moves the fingers to the calculated positions to achieve the specific grasp type.

The observation sub-module re-observes the object at the new position and determines the contact web position. Initially, the sub-module re-observes the object after moving to a new position[1] determined by the find module and recalculates the depth image within the object's bounding box. This depth image is then fed into a Convolutional Neural Network (CNN) specific to each grasp type, which calculates the contact positions of the contact web. As detailed in Sect. 3.2.1 subsection, this CNN extrapolates the visible surface of the object from the depth image and infers the shape of the opposite side to calculate the contact web. Consequently, for multi-finger hands, the contact web may include areas on the backside of the object that are not visible in the depth image, enabling stable grasping. The obtained contact-web position is then passed to the execution sub-module.

The execution sub-module performs the grasp action based on the contact-web position. The execution sub-module starts from the location specified by {approach}, a vector with

---

[1] Depending on the sensor used and the length of the robot's arm, the current implementation uses a distance of approximately 60cm from the object.

a defined length, relative to the object position, and incrementally calculates the positions of the hand and fingers to reach the contact-web positions while approaching the object in the direction of this vector. During this process, similar to human grasping movements, the gripper is expanded wider than its current state at approximately three-quarters of the approach length. Once each fingertip reaches the position in the normal direction of the contact-web, the fingers begin to close following the trained policy.

As for the whole-body posture, it is computed using the role-division algorithm based on the hand position and labanotation, just as in the case of `Bring`.

The termination conditions for grasping vary depending on whether there are force sensors on the fingertips. In the case of the shadow hand used in this instance, the contact resistance can be calculated based on the motor load required to reach the target position of each joint in the fingers. This calculation is utilized to set the termination condition when the contact resistance at the fingertips exceeds a certain threshold. Conversely, for the THK-hand and parallel hand, which rely on passive compliance with simple opening and closing motions, the termination condition is set when a command to close is issued as the fingertips approach the vicinity of the contact-web. It should be noted that the skill agent of the shadow hand employs policies obtained through reinforcement learning.

**`Pick` task model** For the `Pick` task model, the skill parameters include which hand to use, the lifting direction from environmental objects such as a table, and the posture at the beginning of the Pick task.

```
Pick {hand}{departure} {labanotation}
```

The functioning of the `Pick` skill agent closely mirrors that of the `Bring` skill agent. When this agent is activated, the hand is positioned where the previous task concluded, specifically in the state of grasping the object. Consequently, the agent's primary objective is to transition from the current location in the specified direction {departure}. Unlike the Bring task, where the goal position is defined by absolute positional data, the `Pick` skill computes the goal position using this departure vector from the current location. As a result, if the obstacle distribution around the object is roughly the same across both observation and execution phases, the same strategy can be reused. Similar to the `Bring` task, partitioning the initial and goal positions calculates intermediate goals that the robot iteratively follows. Additionally, the overall posture of the robot is determined using the role-division algorithm with Labanotation.

Regarding the termination condition, since {departure} is provided as a vector with an explicit length rather than a unit vector, the `Pick` skill agent ceases operation upon reaching the goal position deduced from the vector's length. This implies that the `Pick` task is endowed with local knowledge to maintain the specified departure direction until the robot moves a sufficient distance from the environmental object, irrespective of trajectory deviations during the demonstration.

**Place & Release task model** The operation of the `Place` skill agent is nearly identical to that of the `Pick` skill agent. However, their termination conditions differ slightly. In the case of `Pick`, the required movement distance is predefined. In contrast, for `Place`, the operation terminates not based on distance, but upon detecting a collision with the environment. Specifically, in the `Place` scenario, displacement is computed from the approach direction, and termination is triggered when a force sensor detects contact between the object and its environment—such a collision serves as the end condition.

During each step, force sensor readings are monitored: if there is no change, the system proceeds to the next step; if the readings increase due to a collision, the system halts. This approach is used because the positions of environmental objects, such as tables, derived from demonstrations, may contain errors—making collision detection a more reliable criterion for termination.

Regarding `Release` skill, after opening the fingers, the hand moves in the direction specified by the departure vector for the indicated distance, similar to the functioning in Pick skill. The final `Bring` task involves the hand returning to its standard position.

```
Place    {hand}{approach}{labanotation}
Release  {object}{grasp-type}{departure}{labanotation}
Bring    {hand}{goal}{labanotation}
```

**Performance** Fig. 2.3 illustrates the execution process. The success rate during execution was approximately 90% across 20 trials. The main cause of failure was the error in the estimated position of the contact web during execution. Another contributing factor to the increased failure rate was instances where the object was too far from the edge of the table,

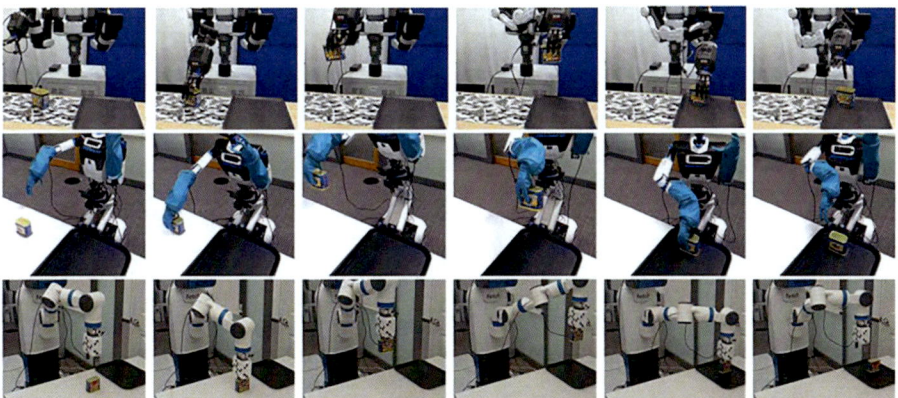

Grasp the box    Pick up the box    Bring the box    Place the box    Release the box

**Fig. 2.3** Tray GMR task-cohesion (Batch mode)

preventing the role-division algorithm from resolving the posture. The position given by the find module is a bit far from the object due to the need to prevent the lower part of the robot from colliding with the desk.

### 2.3.2  Interactive mode (Shelf GMR Task-Cohesion)

In the interactive mode of LfO, it is possible to bypass part of the decoder and provide step-by-step verbal instructions with partial demonstrations to execute the robot sequentially. This can serve as a support function for teleoperation. We refer to this as "symbolic teleoperation."

Traditional teleoperation facilitates the remote control of robots by humans, thereby presenting a practical approach to managing robotic movements. Its direct teaching method, which obviates the need for specialized knowledge, has rendered it widely adopted in various fields. However, conventional teleoperation systems frequently fail to comprehend the context of their actions, leading to the uncritical replication of all human-taught movements. As a result, this may culminate in the execution of erroneous actions or the assimilation of non-essential nuances of the operator's movements. The necessity for sustained vigilance in such systems can induce tension and fatigue in the operator. To address these challenges, it is imperative for teleoperation systems to possess an awareness of their current operational context, thereby enabling them to selectively mimic only the critical aspects of the instructed movements.

LfO can be utilized as a teleoperation support function by enabling the system to focus on specific partial movements rather than all movements based on verbal instructions with task models. This interactive mode, referred to as *Symbolic teleoperation*, significantly enhances the efficiency and safety of remote robotic operations. By employing verbal instructions and task models, LfO facilitates selective imitation, thereby enhancing the overall teleoperation process.

LfO also addresses the drawback of teleoperation, which requires re-teaching the robot each time. Another issue with teleoperation is that the system lacks awareness of its actions. Consequently, demonstrations are necessary whenever new situations arise. LfO retains the knowledge of tasks once performed, along with their affordance information, in task models. By saving these task models, there is an advantage in being able to reuse the previous task models to control the robot in similar situations at a later date. This benefit lies in LfO's capability to symbolically represent movements.

In the interactive mode (symbolic teleoperation), the system is sequentially instructed for each task – such as `grasp` or `pick` – in a sequence of tasks by specifying the task to be learned using verbal instructions, followed by the human operator demonstrating the corresponding action. This verbal input is fed directly into the Task Cohesion Generator to obtain the corresponding task model. Utilizing this task model, the Affordance Analyzer

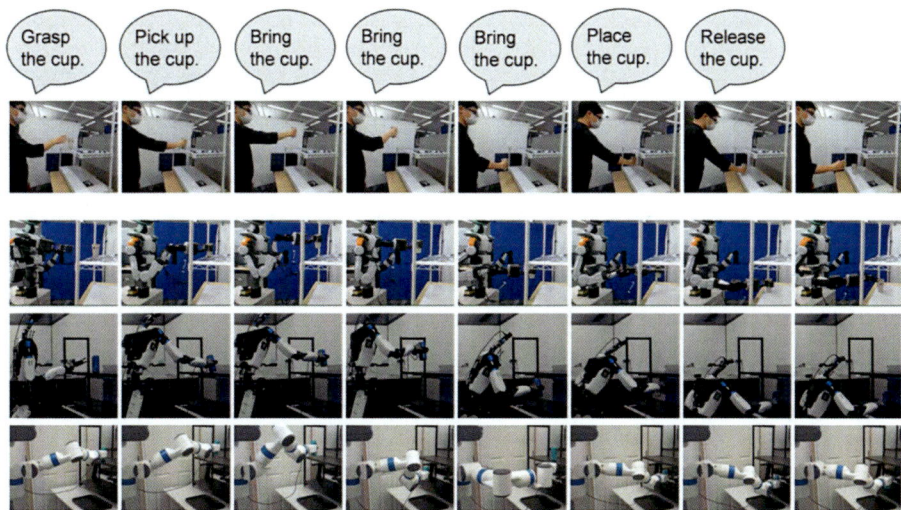

**Fig. 2.4** Shelf GMR task-coheision (Interactive mode (Symbolic teleoperation))

observes only the essential affordance information from the human demonstration. To enter this mode, the operator declares "start teaching" before a series of actions. After demonstrating the sequence of operations using both language and actions, the operator exits the mode by declaring "stop teaching." In other words, the following verbal instructions are sequentially provided, as depicted in the upper section of Fig. 2.4.

```
Start teaching
Bring the hand here
Grasp the cup
Pick the cup
Bring the cup
Bring the cup
Bring the cup
Place the cup
Release the cup
Bring the hand here
Stop teaching
```

The following task models are generated from the verbal input and the demonstration. Then, sequentially, the robot's actions are produced as shown in the lower sections of Fig. 2.4.

```
Bring     {hand}{goal}{labanotation}
Grasp     {object}{hand}{grasp-type}{approach}{labanotation}
Pick      {hand}{departure}{labanotation}
Bring     {hand}{goal}{labanotation}
Bring     {hand}{goal}{labanotation}
Bring     {hand}{goal}{labanotation}
Place     {hand}{approach}{labanotation}
Release   {object}{hand}{grasp-type}{departure}{labanotation}
Bring     {hand}{goal}{labanotation}
```

Consider the advantages of symbolic teleoperation over conventional teleoperation in this context. Firstly, in the `Grasp` task, by declaring "Grasp the cup," the demonstrator does not need to concern themselves with the movements of each finger, as the `Grasp` skill automatically executes the grasp action. Similarly, for the `Pick` task, the system only needs to be instructed on the direction to move away from surrounding environmental objects, disregarding other movements, to effectively perform the `Pick` action. In this task cohesion, `Bring` is used consecutively three times to teach the robot intermediate points to avoid collisions with the shelf. In other words, the human demonstrator plans a collision-free path and uses way-points to teach this collision avoidance path to the system. In this case, only the critical way-points are retained, allowing minor trajectory fluctuations to be ignored and the focus to remain solely on placing these essential points in free space.

## 2.4    Related Work

### 2.4.1    TSS

Numerous simulation environments for robotics research and deployment have been developed, offering valuable tools for advancing robotic technologies [60, 61]. Notable examples include Gazebo [62], MuJoCo [63], CoppeliaSim [64], CARLA [65], AirSim [66], and Webots [67].

Gazebo excels in simulating executions using sensors and actuators integrated with ROS, making it a valuable tool for robotics. It is highly popular within the robotics community due to its compatibility with a wide range of robots. However, its slower simulation performance and inconsistencies in physical simulation render it less ideal as a simulator for machine learning applications.

MuJoCo is well-suited for providing stable physical simulations in machine learning applications, which has made it widely adopted within the machine learning community. However, it lacks certain features essential for robotics simulations, such as inverse kinematics and visual feedback through realistic rendering.

CARLA and AirSim focus heavily on physical simulations in scenarios such as drone operation, offering exceptional visual environments. However, as they were primarily developed for navigation rather than manipulation, they lack crucial features required for robotics manipulation, such as kinematics essential for manipulation tasks.

Many of these tools are highly specialized for simulation tasks, making it somewhat challenging to use them seamlessly across the entire workflow—from simulations designed for machine learning, to those for debugging, and ultimately to real-world robotic execution. When considering robotics machine learning, integrating machine learning frameworks with simulators presents a promising approach.

PyRep toolkit [68] utilizes CoppeliaSim as a simulation environment to establish machine learning frameworks. Similarly, Deepbots [69] employs Webots as a simulator, enabling external machine learning functionalities developed within the community. However, these environments are not designed to seamlessly bridge the gap between training and execution. Instead, training and execution occur as separate processes, making the transition from learning to real-world implementation less fluid.

The task sequencing simulator introduced in this chapter orchestrates multiple simulators, bridging simulations designed for learning with those aimed at execution, thereby enabling a seamless workflow from training to implementation. Moreover, the simulator can be utilized in various stages—whether pre-trained, in the midst of training, or during execution. For example, it can be integrated as a plugin for machine learning platforms and subsequently connected for execution on ROS, facilitating a streamlined transition from simulation to real-world application.

## 2.4.2   Role Division Mapping

In the field of robotics, various methodologies have been developed to translate human movements into robotic actions. These include directly mapping human body motions to robot joint angles [70] and kinesthetic teaching, where robot joints are passively manipulated for teaching purposes [4]. Recent advancements have introduced approaches that combine kinesthetic teaching with corrective adjustments [71].

Many of these methods involve modeling the trajectories observed during demonstrations through mathematical frameworks, such as Hidden Markov Models, Gaussian Mixture Models, or combinations thereof [72]. Additionally, these trajectories can be mapped using dynamic equations or probabilistic formulations to enhance the accuracy and reliability of robotic behavior [73, 74].

These direct mapping methods become increasingly challenging when manipulation actions require base movements of a robot. Few studies have tackled the issue of mapping or automatically implementing base movements from demonstrations.

A notable exception is the work of Welschehold et al., who addressed this problem by mapping human torso movements for single-arm tasks [75]. Furthermore, they proposed a method to learn individual actions and resolve ambiguities in overall task goals within mobile environments using a limited number of demonstrations [76]. One limitation of these approaches, however, is the need to repeat demonstrations multiple times [3].

Achieving comprehensive knowledge and understanding of tasks and intentions solely from demonstrations remains an ambitious goal. Akgun et al. [77] proposed explicitly teaching constraints to ensure accurate instruction by humans, thereby avoiding mismatches with mental models. Their research focuses on task symbolization and geometric constraints. Early work in this direction includes studies emphasizing object state transitions [9]. More recently, Perez et al. [78] introduced multi-step task demonstrations that combine geometric constraints and keyframes.

Most of these approaches concentrate on end-effector movements, assuming that task constraints can be resolved using general motion planners. However, this assumption holds only for robots with sufficient reachability and is not applicable to compactly structured domestic robots [79]. For such domestic robots, accomplishing complex task sequences requires precise full-body positioning and integrated arm motions. Addressing this issue necessitates resolving motion constraints arising throughout the entire sequence.

The method proposed in this chapter utilizes demonstrated motions to implicitly resolve constraints. By combining task-specific knowledge—such as instructions on what to manipulate—it eliminates the need to repeatedly perform demonstrations for learning the teacher's intent. Moreover, integrating motion knowledge derived from a single demonstration allows for planning appropriate motions within a task sequence without the necessity of modeling all ambiguities using expert knowledge. This approach streamlines the process and reduces reliance on exhaustive demonstrations, paving the way for efficient task execution in robotics.

In terms of combining task and motion knowledge, recent studies aim to utilize language as a medium to integrate task instructions with demonstrations performed through teleoperation [80]. However, the reliance on teleoperation poses significant challenges, particularly when teaching dual-arm robots or mobile-based robots, highlighting the complexity of effectively conveying instructions in such scenarios.

Methods for mapping human full-body movements to robots are not limited to the robotics field discussed here; they are also applied extensively in areas such as computer graphics [81–85] and machine learning [86–89] communities.

However, these approaches in other fields primarily focus on replicating realistic movements in robots or avatars, emphasizing the appearance of the motion itself. They do not aim to address external forces, such as resistance from the environment, to ensure task success. Consequently, when these methods are applied to actual robots, they face the challenge of low success rates in completing tasks due to their limited consideration of environmental dynamics and operational constraints. This presents a significant limitation when transitioning such techniques to real-world robotic applications.

# Grasp-Skill Library and Its Design

**3**

This chapter and next chapter detail the design of skill agents employed by the decoder. During the design phase, which precedes the execution phase, skill agents are meticulously prepared for each task.

Tasks and their corresponding skills are selected based on their necessity and sufficiency. As previously mentioned, a task is a term that expresses a single unit of robotic action, such as picking up or grasping, according to its purpose. A skill refers to a set of control commands used to execute that action with a specific robot. When the robot is fixed, tasks and skills correspond one-to-one. However, there is a difference between viewing the action purposefully, as denoted by the term "task," and viewing it procedurally, as denoted by the term "skill."

Robotic tasks are classified into two primary categories: grasping tasks, which involve the robot's hand making contact with an object, and manipulation tasks, where the object is already in the robot's hand and moves relative to the environment due to the hand's motion. This chapter covers the design of the grasp tasks and corresponding skills, while the next chapter covers the design of the manipulation tasks and corresponding skills. The necessary task sets are established using Closure theory [17] for grasp tasks and Kuhn-Tucker theory [18] for manipulation tasks. Corresponding skills for these necessary task sets are compiled in the grasp skill library and the manipulation skill library.

© The Author(s), under exclusive license to Springer Nature Switzerland AG 2026       33
K. Ikeuchi et al., *Learning-from-Observation 2.0*, Synthesis Lectures on Computer Vision,
https://doi.org/10.1007/978-3-032-03445-8_3

## 3.1    Grasp Taxonomy and Closure Theory

Humans are known to employ various grasp methods for the same object depending on the purpose of the action and the situation, as illustrated in Fig. 3.1. To grasp a pen firmly, a method with a larger contact area between the hand and the object, as shown in Fig. 3.1a, is used to increase resistance. Conversely, to point with a pen, a method that allows for subtle directional control using the remaining degrees of freedom of the finger joints, as depicted in Fig. 3.1b, is employed. When writing, a method that enables appropriate force application and precise directional control of the pen is used. This method utilizes the area between fingers as a pivot point to constrain only the translational motion of the pen while controlling its direction with the fingertips, as shown in Fig. 3.1c.

To analyze these grasp methods and use them as references for robotic grasping actions, various classifications of human grasping methods have been proposed in the field of robotics [90–92]. One of the pioneering works in this field is Cutkosky's classification, which attempted to build a taxonomy of human grasp methods and use it through rule-based approaches to determine the robot's grasp method. However, challenges remained. Using Cutkosky's taxonomy as a starting point for discussion, Kang et al. proposed a representation method called the contact web, which represents the distribution of contact points between parts of the hand and the object. They classified human grasp methods based on the contact-web pattern, as shown in Fig. 3.2. Their taxonomy first classified grasp methods based on whether the palm was involved in the contact web, and then further classified them by the spatial distribution of the thumb's contact points, represented by squares in the figure, and the contact points of the other fingers, represented by black dots in the figure. Using the contact webs, they observed human grasp methods and mapped them to the grasp methods of the Uhta-MIT robot hand. However, directly mapping grasp methods led to an increase in the number of skills to be implemented and issues with horizontal transfer to

(a)                            (b)                            (c)

**Fig. 3.1** Grasping a pen

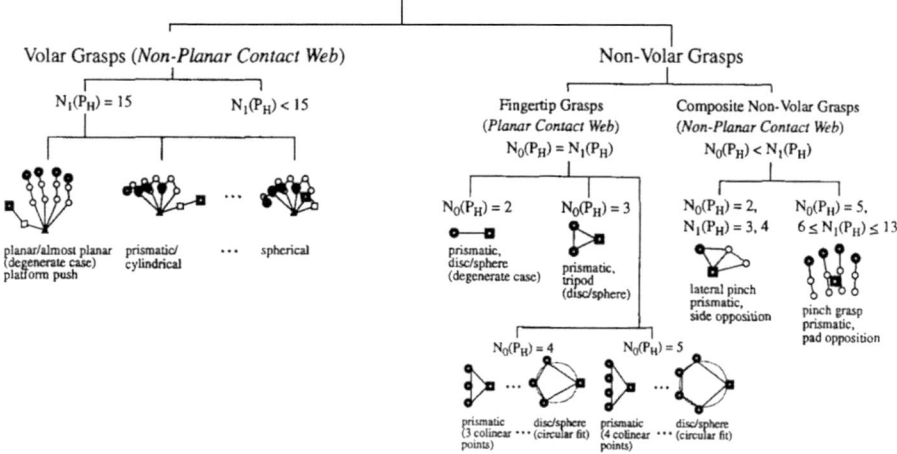

**Fig. 3.2** Kang et al.'s grasp taxonomy based on contact-web representations [91]

other hardware. In the current skill implementation, we use Kang et al.'s contact web as an intermediate representation.

Grasp methods are categorized according to the intended purposes of human actions. When replicating human-demonstrated grasps, it is inefficient to mimic them exactly due to the structural differences between human and robotic hands. Consequently, our approach emphasizes understanding the underlying purposes of these human grasp methods and adapting the robot's grasp techniques accordingly.

Yoshikawa et al. categorized grasp methods performed by both humans and robots according to their purposes and subsequently introduced three distinct types of closure [17]:

- **Passive-force closure**: This closure involves applying pressure to secure an object at a specific point in space, similar to a vise. It utilizes nearly all the degrees of freedom of the fingers and hand, ensuring contact with the object and exerting pressure to hold it in place. Consequently, the relative relationship between the object and the hand is fixed. Typically, the contact-web points are distributed in multiple directions, often forming a convex polyhedral shape. In Kang et al.'s classification, this corresponds roughly to "Volar Grasps," and in Felix's classification, it aligns with "Power Grasp."

- **Active-force closure**: This closure involves holding an object with minimal contact, similar to using tweezers, while controlling its orientation. By utilizing some degrees of freedom of the hand and fingers, the object is grasped, while the remaining degrees of freedom allow for control of its orientation relative to the hand. This closure is also often employed to pick up a small object from a narrow, cluttered area before in-hand manipulation in the air to settle into the final grasp. Generally, the contact-web points are aligned on a single plane. In Kang et al.'s classification, this roughly corresponds to

"Fingertip Grasp" within Non-Volar Grasp, and in Felix's classification, it aligns with "Precision Grasp."

- **Passive-form closure**: This closure involves constraining the position and orientation of an object to some extent while allowing movement within that range, similar to a wheel bearing. The fingers and hand form a confined space around the object, limiting its position and orientation within that space. Depending on the object's position within the allowable range, the distribution of contact-web varies. In Kang et al.'s classification, this category does not exist because its contact-web does not form. Similarly, there is no clear corresponding category in Felix's classification. Among the grasps recognized as corresponding to Passive-force closure, the decoder projects those that require this to this category, taking into account task cohesion, such as grasp in open-a-refrigerator-door cohesion.

## 3.2    Grasp Skill Agent and Its Design

The three closures will serve as the fundamental principles for designing the grasp library. Each entity that executes a grasp skill is referred to as a grasp skill agent. Upon activation by the decoder, each grasp skill agent autonomously performs its respective grasp skill by utilizing its vision module to gather environmental information and executing the skill based on policies obtained through reinforcement learning. Three agents are designed to correspond to the three closures, passive-force, active-force and passive-form.

Each skill agent can be divided into observation and execution modules. The observation module calculates the contact web for a closure from the input depth image by utilizing a Convolutional Neural Network (CNN). The execution module forms the closure based on force feedback and other factors, targeting the positions of the contact web. The execution module is pre-trained using reinforcement learning. See Fig. 3.3.

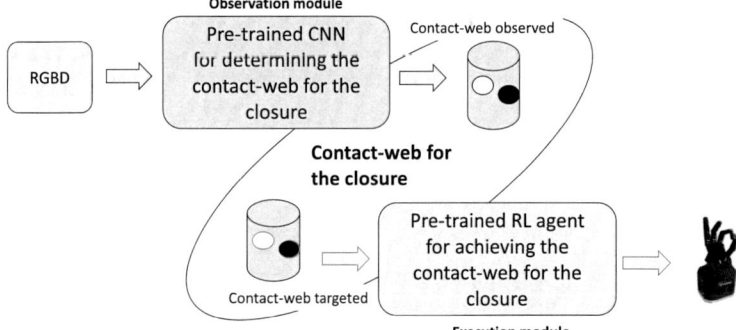

**Fig. 3.3** Contact-web based visuomotor pipeline

### 3.2.1  Observation Module in an Agent

The observation module utilizes a CNN capable of calculating a specific contact web for a closure, regardless of the shape or size of the object. For grasping with a parallel gripper, contact points can only be placed along the object's visible contour. In contrast, with a multi-fingered hand, stable passive force closure can only be achieved by placing contact points on both the visible and unseen sides. In such cases, the CNN is trained through supervised learning to extrapolate and determine contact points on the opposite, unseen side from the visible data, in addition to the visible side.

The training data for the CNN is obtained by assuming that all objects can be approximated by superquadric. One depth image is obtained from a superquadric equation and the sensor's direction. The entire surface of the superquadric is used to calculate the contact web, creating one pair of training data. This process is repeated while varying the sensor direction and superquadric's parameters to create the training data.

A superquadric can be expressed by a mathematical equation in Eq. 3.1. While the exponent in an ellipsoid is 2, in a superquadric it is a real number. These exponents, $\epsilon_1, \epsilon_2, \epsilon_3$, are called the shape parameters. Figure 3.4 illustrates examples of shape changes when $\epsilon_1, \epsilon_2, \epsilon_3$ are varied. On the other hand, the sizes of both ellipsoids and superquadrics can be changed by altering the denominator parameters, $a, b, c$, referred to as size parameters.

$$\left(\frac{x}{a}\right)^{\epsilon_1} + \left(\frac{y}{b}\right)^{\epsilon_2} + \left(\frac{z}{c}\right)^{\epsilon_3} = 0. \tag{3.1}$$

Depth images for training the CNN are obtained using superquadrics. The visible surface by a sensor should be generated according to the sensor model [93] of the sensor being used. For example, with a light-stripe sensor that irradiates laser light from an oblique direction, shadow parts visible from the camera cause the visible surface to skew to one side. In the case of photometric stereo, only the common parts illuminated by light from three directions are visible. For the binocular stereo vision sensor used in this study, depth information is only observed from the surfaces visible to both cameras. By assuming that the distance from

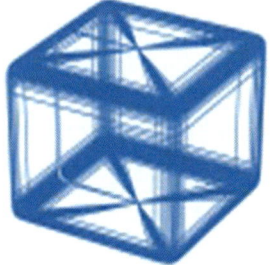 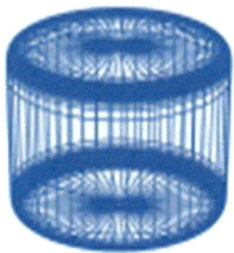 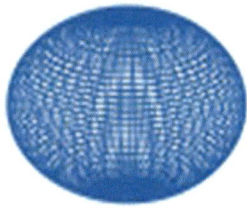

**Fig. 3.4** Superquadric surface. Shape variations of objects can be expressed through changes in shape parameters

the sensor is significantly larger compared to the distance between the cameras, the visibility of a point is determined based on the angle between the normal direction and the assumed line of sight. The depth distribution of only visible points is taken as the output from the sensor.

The location of the contact web is analytically calculated from the superquadric equation with respect to the shape of the gripper used. The contact web varies depending on the gripper used. Here, we explain the case of the four-fingered Shadow hand as an example.

In the case of passive force, as shown in Fig. 3.5a, the approach direction is determined by the azimuthal angle of the line of sight in the XY plane of the superquadric. The gripper is moved towards the center of the object until the palm part makes contact, then the fingers are closed until they contact the object, and these contact points along with the initial palm contact point form the contact web. For the passive form, the procedure is similar to that of the passive force, but the fingers are closed halfway after palm contact is established, with the fingertip positions forming the virtual contact web.

For active force contact web, as illustrated in Fig. 3.5b and c, two scenarios are considered: one where the z-axis of the superquadric aligns with the direction of gravity, and another where it is perpendicular to gravity. The former corresponds to a cylindrical object standing upright on a table, while the latter corresponds to one lying flat. In the first scenario, the z-axis of the superquadric serves as the approach direction, and the fingertips are positioned on a plane–referred to as the closure plane–located at a certain distance from the tip of the superquadric and orthogonal to the z-axis. In the second scenario, the closure plane is defined as the plane that contains the superquadric's z-axis and is orthogonal to gravity; the fingertips are then arranged along the contour of this plane.

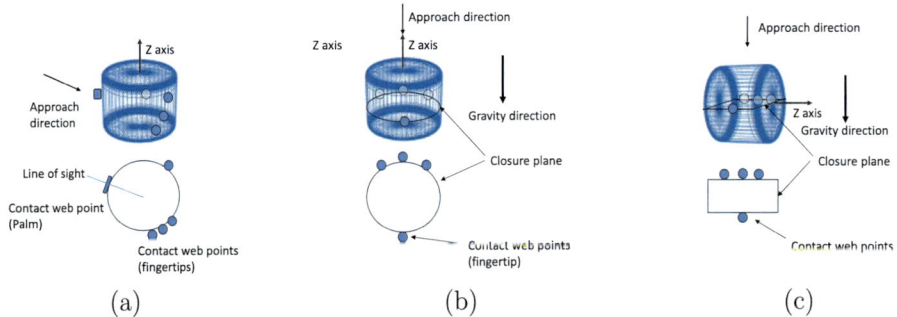

**Fig. 3.5** Example of contact web computation. The structure of the contact web depends on the hand being used. This example is based on a four-finger Shadow Hand. **a** In the passive force case, the palm first contacts the object surface from the approach direction, followed by the other fingers wrapping around the object to establish contact. **b** Active force case 1. When the Z-axis of a superquadric is the approach direction, contact web points are arranged on the closure plane perpendicular to that direction. **c** Active force case 2. Contact web points are configured along the closure plane that contains the superquadric's Z-axis and is orthogonal to gravity

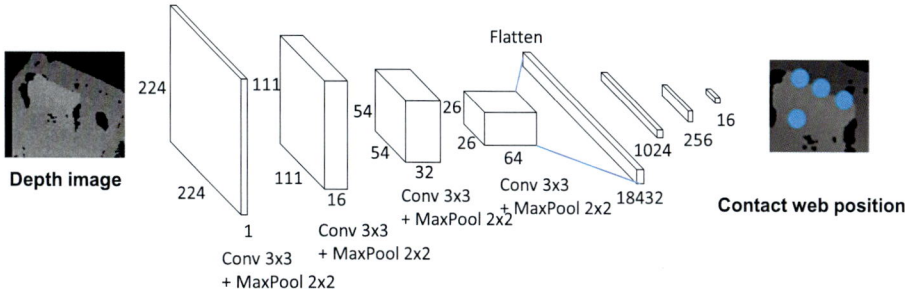

**Fig. 3.6** CNN for observation module. The input is a depth image, and the CNN consists of four convolutional layers and four fully connected layers. The final layer is 16-dimensional vector, representing the 3D positions of contact webs as well as a quaternion that describes the overall posture of the hand

A depth image and contact web position are obtained from the superquadrics equation by randomly sampling the size parameter and shape parameter, creating one training data point. Repeating this process multiple times provides the training dataset for the Observation module's CNN.

The CNN consists of four convolutional layers and three fully connected layers. These layers are trained with 400,000 pairs of true contact webs and depth images, and they output the position of the contact web and the object's centroid. See Fig. 3.6.

The resulting CNN will be capable of calculating the contact web from the input depth image. Note that the CNN extrapolates the unseen opposite side's shape from the assumed superquadric based on a single-direction depth image and estimates the contact web.

### 3.2.2   Execution Module in an Agent

The execution module in a skill agent is trained through reinforcement learning. Pybullet and the Proximal policy optimization (PPO) algorithm are utilized as the simulator and the training algorithm, respectively. The state variables provided to the algorithm include the robot's fingertip positions, wrist orientation, and the reaction forces on the fingertips. The center position of the object and the approach direction are derived from human demonstrations and serve as hint information. The initial values of the joint angles are set to their neutral positions, and the wrist position of the robot is established at a predefined distance along the approach direction. The state variables are updated at regular intervals, with the action values representing the subsequent joint angles of the fingers and the next wrist posture.

The reward function  is common to three types of closures, differing only in the positions of the target contact-web. The locations of the target contact-web are calculated using the superquadric equation, whose size and shape parameters are randomly set for training, for a particular closure. For each contact-web point $i$, the target contact position $\mathbf{c}_i$ and the target contact direction $\mathbf{n}_i$ are calculated. Let $N$ be the number of contact-web points. At each time step $t$, the distance $d_{i,t}^p$ between the current position $\mathbf{p}_{i,t}$ of the robot's fingertip and the target contact-web position $\mathbf{c}i$, and the cosine angle distance $d_{i,t}^f$ between the fingertip direction $\mathbf{f}_{i,t}$ and the normal direction $\mathbf{n}_i$ at that contact-web point, are calculated. Here, the fingertip direction is calculated from the direction of the robot's hand and finger joint axes, and if the finger is not in contact with the object, $d_{i,t}^f$ is defined as 0. Using these and the upper limit values $b^f$ and $b^b$ as references, the reward $r_t$ is defined as follows.

$$r_t = \sum_{i=1}^{N} \log (b^p - d_{i,t}^p) + \log (b^f - d_{i,t}^f), \tag{3.2}$$

where

$$d_{i,t}^p = \|\mathbf{p}_{i,t} - \mathbf{c}_i\|,$$
$$d_{i,t}^f = \arccos(\mathbf{f}_{i,t} \cdot \mathbf{n}_i).$$

The performance of the active/passive closure is assessed to ascertain whether sufficient contact force has been generated to counteract external forces, such as gravity at the end of an episode. Should this test fail, a penalty is imposed, which is then factored into the final reward calculation.

During training, curriculum learning was applied with randomization of both shape and size parameters. Initially, the shape and size parameters were fixed to allow the agent to learn grasping relatively easily. Once the convergence rate saturated, the size parameters were randomized and training resumed. In the third stage, shape parameters were also randomized for further learning. In the fourth stage, the reward weight was increased to encourage the agent to adopt a more stable posture when lifting objects.

The figures illustrate actions performed by the trained agent. Figure 3.7a shows an example from Active Force Scenario 1, where the same agent successfully grasps objects of two different shapes. Figure 3.7b demonstrates an agent using passive force to grasp the same object. Figure 3.7c depicts a door-opening task, where the agent uses the passive form to secure a door handle within the area enclosed by the fingers.

## 3.3    From Demonstration to Grasp Skill Execution

### 3.3.1    Mapping Table

The grasp pipeline consists of three agents corresponding to the three closure types, while it is challenging to derive these closures directly from human demonstrations. Conversely, the use of grasp taxonomy to recognize human grasps is prevalent. To address this, we meticulously examined the taxonomy proposed by Kang et al. and constructed a correspondence table,

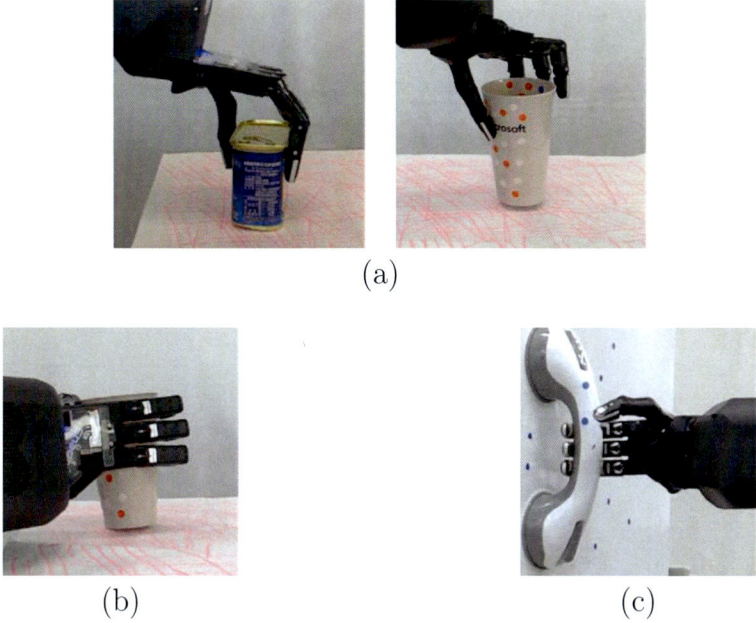

**Fig. 3.7** Grasp skill agents trained. **a** Active force grasp agent. Objects of different shapes are grasped by the same active force agent. **b** Passive force grasp agent. The cup on the right in (a) is grasped using a different way by the passive force agent. **c** Passive form grasp agent. In the case of the passive form grasp, the target object can move within the hand. Note that the thumb is positioned to enclose the region, but does not make contact

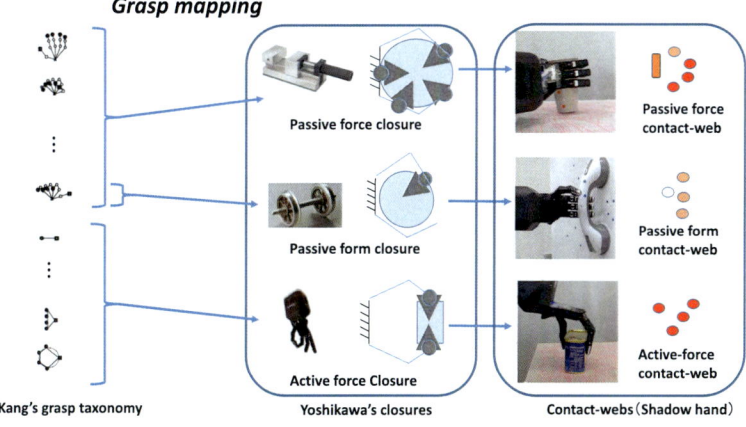

**Fig. 3.8** Grasp mapping from taxonomy to closure

as depicted in Fig. 3.8. In this figure, the left column represents Kang et al.'s taxonomy, the center illustrates the three closure types, and the right column provides examples using the THK hand. It is noteworthy that certain grasps apply to both passive force and passive form. The decoder assesses cohesion to determine the appropriate application.

## 3.3.2   Online Control Flow

Figure 3.9 illustrates the control flow in which grasp methods are recognized, mapped to closures, and executed. During the teaching phase, the demonstrated human grasp is identified by the grasp recognition system [94], which is based on Kang's classification. This yields one grasp type from Kang's classification as the recognition output. Using the predefined mapping table from Kang's grasp types to closures, one of three candidate closures is selected. The affordance analyzer then registers the selected closure within the task model. In the execution phase, the decoder activates the corresponding skill agent to the closure.

During the execution phase, the grasp skill agent corresponding to the closure is activated along with affordance information from the task model. Figure 3.10 illustrates the activity of a grasp skill agent. The activated agent utilizes the object's name to obtain the bounding box of the object region in the RGB image using an off-the-self vision module. It then extracts the corresponding segment in the depth image and inputs it into the CNN, which outputs the positions of the contact web and the object's centroid. With this contact web information and object centroid serving as initial values, alongside the approach direction, the agent determines the actions of each finger based on the policy derived from reinforcement

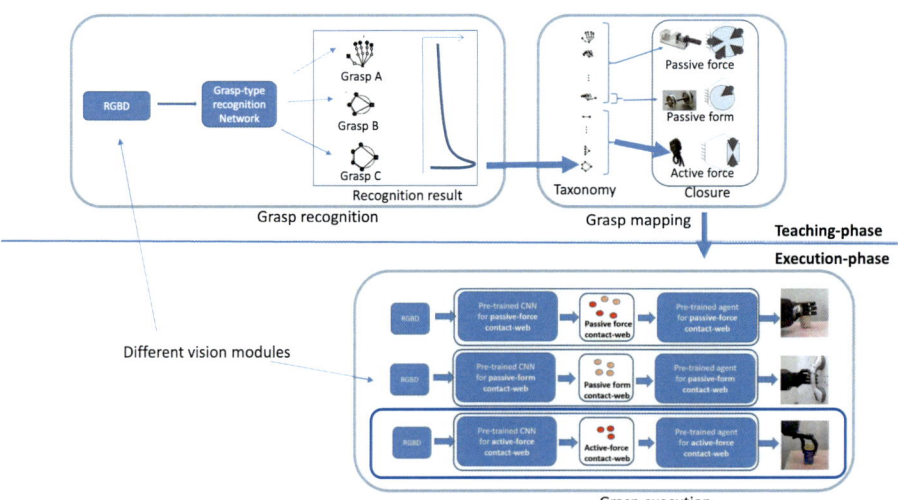

**Fig. 3.9** Grasp execution flow

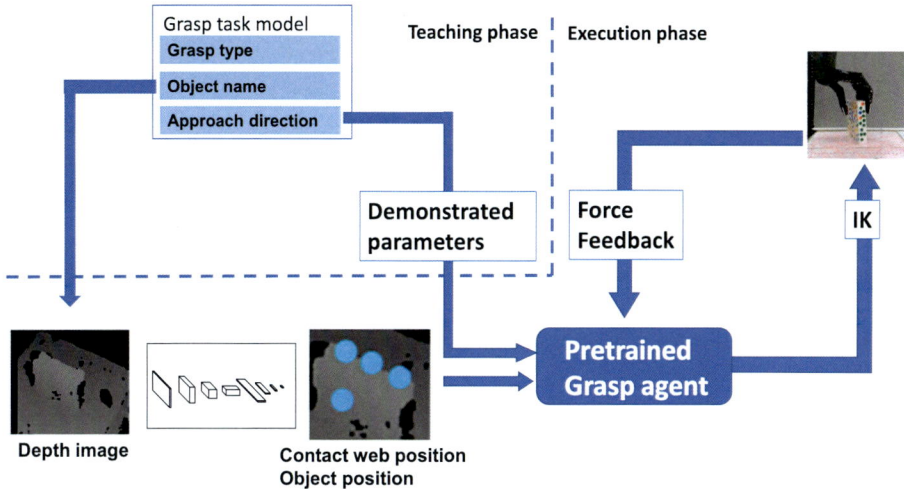

**Fig. 3.10** Grasp skill agent at the execution phase

learning. The arm motion is generated through IK to achieve these finger positions. The series of loops continues, integrating force feedback information, until the termination conditions are met, at which point the agent's execution concludes.

## 3.4   Related Work

### 3.4.1   Stable Grasp

Grasping refers to the process in which a part of a robot's body, typically a robotic finger, makes contact with an object, secures it to a part of the hand, and prepares for subsequent tasks. In the field of grasping research, a key challenge lies in the formulation of stable grasping, which involves defining and maintaining the relationship between the object and the robotic hand [95, 96]. Many approaches presume prior knowledge of the model of the object to be grasped when calculating grasping configurations [97–103]. For example, in the GraspIt! [104] simulations are employed to assess grasp stability indicators [100] under the assumption that the object's shape is already known. These simulations determine optimal finger arrangements, which are subsequently utilized in motion planning to achieve the desired grasp configuration.

After determining a stable grasping strategy, in a cluttered environment, a path is planned to approach while avoiding surrounding objects [95, 97, 99]. Rather than simply avoiding obstacles, some research focuses on planning a grasping strategy while moving objects [103]. In any case, these studies assume that the shape of the objects is known in advance.

With advancements in deep neural network technologies, several approaches leveraging convolutional neural networks have been proposed for detecting grasp points on objects whose models are not pre-defined [105–107]. These methods typically identify grasp points by extracting features from images, under the assumption that stable grasping can be achieved when certain relationships, such as a pair of parallel lines, among these features are satisfied. Nevertheless, these early techniques, which rely on open-loop control, are notably vulnerable to inaccuracies in the estimated pose of the object.

To address this limitation, several end-to-end systems have been proposed that control the gripper with visual feedback, enabling stable grasping [108–110]. Additionally, multimodal grasping methods that integrate visual and tactile inputs have been developed [111, 112], showing great promise in achieving stable grasping. Nevertheless, many of these methods simplify the problem by employing parallel-jaw grippers or by focusing solely on achieving a stable grasp without considering the contextual requirements of the task.

### 3.4.2   Grasp Taxonomy

Alongside research focused on stable grasping, significant efforts have been devoted to classifying grasping methods into various categories from diverse perspectives. The objective of these studies is to achieve task-oriented grasping [48], which involves determining the most suitable grasp for a specific task or a task sequence. Examples include the classification of human grasping strategies based on object shapes and the intended purpose of the action [90, 92, 113–116], as well as the categorization of robotic grasping methods from the viewpoints of stability and manipulability [17].

Building upon established classifications of grasping, researchers have explored methods for determining appropriate grasping strategies. Notably, Cutkosky's pioneering work proposed a rule-based approach to select suitable grasping strategies from these classifications by leveraging the characteristics of tasks and target objects [90]. Kang et al. developed a methodology to identify optimal grasping strategies derived from human demonstrations [114]. Moreover, Kang et al. addressed the structural differences between human and robotic hands through the introduction of functional mapping [91]. This approach identifies the functionality of individual fingers based on the virtual finger theory [117] and establishes correspondences between human and robotic fingers, thereby mitigating structural disparities.

However, despite these advancements, the practical implementation of grasp mapping on robotic systems remains ambiguous, primarily due to the intrinsic structural differences between human and robotic hands. As a result, direct replication of human grasping techniques proves challenging, underscoring the incompatibility of applying human grasping classifications directly to robotic systems.

Reinforcement learning methods have also been proposed to pre-train robotic grasping strategies based on various human grasp types. During execution, these strategies are selected according to the specific requirements of the given task [118, 119]. However, certain human grasp types involve redundancies that are unnecessary for robotic applications and can potentially be reduced from the perspectives of stability and manipulability [17]. For example, grasps such as the Prismatic-3 finger (involving four fingers) and the Prismatic-4 finger (involving five fingers) are considered equivalent in terms of stability and manipulability.

Nevertheless, these methods often overlook such redundancies, designing strategies for all possible grasp types, which makes the process more labor-intensive. Furthermore, the reward designs employed in these approaches often disregard contact point positions–a critical feature characterizing grasps–resulting in no assurance that the learned strategies will achieve the desired grasp. This limitation is particularly problematic for tasks requiring the fundamental aspects of grasping, such as appropriate force distribution, to be effectively addressed.

### 3.4.3 Task Grasp

Task-oriented grasping typically refers to methodologies designed to estimate optimal grasping positions on an object, tailored to the specific requirements of a given task [48, 120–130].

These studies focus on estimating grasping positions and contact distributions that enable the object to fulfill its functional role for a specific task, and performing grasps that align with these estimations. For instance, [131] estimates task-oriented grasping positions, while [120] employs reinforcement learning to acquire skills that satisfy contact distributions, ensuring the object can fulfill its function.

However, these approaches primarily emphasize determining where an object should be grasped, neglecting the selection of appropriate grasping primitives. Consequently, there is no guarantee that essential requirements such as the functionality of fingertips or proper force distribution, as discussed in [17, 92], will be adequately addressed.

The core principle of task-oriented grasping involves generating force distributions that align with the task's intent while leveraging the functional capabilities of the fingers. However, the process of mapping human grasping intentions–specifically considering force distributions and finger functionalities–into robotic grasping strategies remains largely unexplored.

Our approach seeks to address this gap by incorporating recent advancements in policy learning methodologies. The key focus lies in mapping human grasping into robotic grasping strategies that transcend reliance on grasping forms alone. Instead, these strategies are designed to optimize force distributions and finger functionalities, ensuring advantages for subsequent tasks. As a future direction, it will also be necessary to consider regrasping an object after picking it up in a way that benefits subsequent tasks [132].

# Mainpulation-Skill Library and Its Design

**4**

The design of the manipulation skill library adheres to the definition of manipulation tasks that cause transitions in surface contact, as established in earlier research [9].[1]

The necessary task sets are established using Kuhn-Tucker theory [18] for manipulation tasks. However, the actual necessity and sufficiency of these skill libraries depend on the specific domain in which the robot tasks are performed. The theories provide an upper boundary for all the necessary tasks; however, in practice, only a subset of these tasks frequently appear in each domain, while the remainder rarely occur. Consequently, it is impractical to prepare all the skills in the libraries. Instead, in the LfO2.0 project, we target the domain of household service robots, designing a comprehensive set of non-overlapping tasks and their corresponding skill agents.

The physical manipulation library compiles physical manipulation skills that effectuate actual physical state transitions. Alongside this, a semantic manipulation library is conceived, which amasses semantic manipulation skills. These semantic skills encompass not only physical constraints but also semantic constraints derived from human common sense to effectively achieve the tasks.

---

[1] As previously described, the term "tasks" is used to conceptualize units of robotic actions based on their purposes, while "skills" refer to the actual methods of execution using a physical embodiment, such as a robot. These terms are used on a one-to-one basis.

© The Author(s), under exclusive license to Springer Nature Switzerland AG 2026     47
K. Ikeuchi et al., *Learning-from-Observation 2.0*, Synthesis Lectures on Computer Vision,
https://doi.org/10.1007/978-3-032-03445-8_4

## 4.1    Physical Manipulation

The physical manipulation library consists of skill agents that generate units of actions to achieve intended tasks based on physical constraints from the environment. These agents assume that the object is already grasped by the hand, and that the motion constraints from the environment on the object are identical to those on the robot's hand.

### 4.1.1    Contact State

The state of the manipulated object (i.e., the state of the robot's hand) is defined by the feasible motion region of the object. The movement of the manipulated object is constrained by the environment at each contact surface. For instance, in Fig. 4.1, the motion region of an object placed on a table, constrained by contact with the table surface, can be expressed as follows:

$$\mathbf{N} \cdot \mathbf{T} \geq 0, \tag{4.1}$$

where $\mathbf{N}$ represents the normal direction of the table surface and $\mathbf{T}$ indicates one possible motion direction of the object. According to the constraint equation, possible motions occur only in directions where the inner product with the table surface's normal direction is positive.

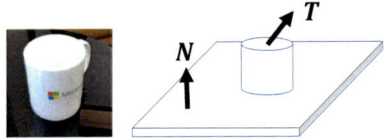

(a) A cup on the table can only move upward.

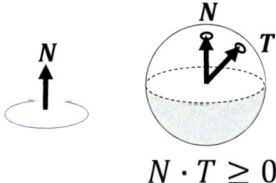

$$N \cdot T \geq 0$$

(b) Possible motions occur in directions where the inner product with the normal to the contact surface is non-negative. Specifically, these directions correspond to the northern hemisphere of the Gaussian sphere, with the north pole representing the direction of the surface normal.

**Fig. 4.1**  One-directional contact

Specifically, motions are constrained to the upward directions relative to the horizontal table surface, while downward motions are inhibited by the table surface.

The right diagram in Fig. 4.1b illustrates the solution area of the constraint equation depicted on the Gaussian sphere. The Gaussian sphere represents the directions of unit vectors as points on the spherical surface, with the starting points of the vectors at the center of the sphere and their endpoints on the surface. Without loss of generality, we can represent the unit surface normal of the table surface as the north pole of the Gaussian sphere. The northern hemisphere, indicating the possible directions, is shown in white, while the southern hemisphere, representing prohibited directions, is shaded gray. The equator represents the constraint where the equality holds, orthogonal to the normal direction, corresponding to a singular region where motion occurs along the table surface.

In multi-directional contact state, involving many constraint directions, the object is further constrained by the normal directions at each additional contact surface, denoted as $\mathbf{N}_1, \mathbf{N}_2, \ldots, \mathbf{N}_n$. Consequently, the range of motion of the object is determined by the solution region of a system of linear inequalities, which is formed by the constraints provided at each contact surface. See Fig. 4.2.

$$\mathbf{N}_1 \cdot \mathbf{T} \geq 0, \tag{4.2}$$

$$\mathbf{N}_2 \cdot \mathbf{T} \geq 0, \tag{4.3}$$

$$\cdots \tag{4.4}$$

$$\mathbf{N}_n \cdot \mathbf{T} \geq 0. \tag{4.5}$$

In other words, with each constraint applied, the solution region (white area) on the Gaussian sphere decreases.

According to the Kuhn-Tucker theory [133], the solution space of linear inequalities is classified into ten types. By applying this general rule to our case, as described in Sect. 6.1, we can classify the solution space of the aforementioned system of linear inequalities into ten types of spherical regions on the Gaussian sphere: the entire spherical surface, hemisphere, crescent region, spherical polygon region, entire great circle, hemi-great circle, segment of a great circle, two points, one point, and none. See Fig. 4.3.

**Fig. 4.2** Multi directional contact

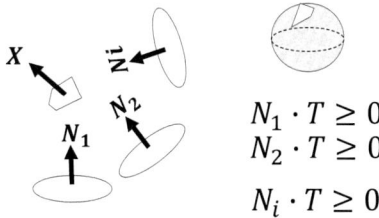

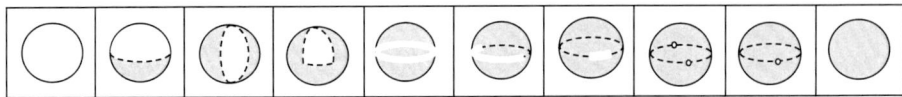

**Fig. 4.3** Ten solution types based on the Kuhn-Tucker theory

The above discussion pertains to the constraints on the translational motion of an object, expressed as a system of linear inequalities in Eq. 4.2. According to the Screw theory [134], similar constraints can be described as follows:

$$\mathbf{N} \cdot \mathbf{T} + p\,(\mathbf{P} \times \mathbf{N}) \cdot \mathbf{S} \geq 0, \tag{4.6}$$

where $\mathbf{P}$ is a vector from the screw position to the contact point, $\mathbf{N}$ is the normal direction at the contact point, and $\mathbf{S}$ is the screw axis vector. A translational motion occurs along $\mathbf{S}$, and a rotational motion occurs around $\mathbf{S}$. The variable $p$ is the ratio between the translation and rotation. For pure translation, $p = 0$, and for pure rotation, $p = \infty$.

If we assume a pure rotation, we obtain:

$$(\mathbf{P} \times \mathbf{N}) \cdot \mathbf{S} \geq 0, \tag{4.7}$$

Here, we denote $\mathbf{M} = \mathbf{P} \times \mathbf{N}$. We obtain

$$\mathbf{M} \cdot \mathbf{S} \geq 0. \tag{4.8}$$

From this, as was the case in pure translation, we obtain 10 solution types for $\mathbf{S}$. In other words, for polyhedral states, ten types of states can be defined for both translational and rotational motion.

Considering their dimensions, we group the ten types into seven states: 3, 2.5, 2, 1.5, 1, 0.5, and 0 DOFs, as shown in Fig. 4.4. Henceforth, we refer to these 14 states–7 states for translation and 7 states for rotation–as the object states or, more precisely, the contact states of the grasped object.

### 4.1.2  State Transition and Task

A physical manipulation task is defined as a unit of robotic actions that causes transitions in the contact states of an object. For example, consider a pick task where an object placed on a table is lifted. Before the `Pick` task, the object's range of motion is constrained to the northern hemisphere of the Gaussian sphere due to the table surface's constraint, i.e., in the `PC` contact state in Fig. 4.5. After the `Pick` task, the object is no longer constrained by the table surface and can move in all directions, with the entire Gaussian sphere representing its range of motion, i.e., in the `NC` contact state in Fig. 4.5. Thus, the purpose of the `Pick` task

| State name | DOFs | Admissible translation directions on the Gaussian sphere | Examples |
|---|---|---|---|
| **NC** Non-contact translation | 3 | NC (3, 0, 0) | Object floating in the air |
| **PC** Partial contact translation | 2.5 | PC1 (2, 1, 0)  PC2 (1, 2, 0)  PCN (0, 3, 0) | Object that can move upward |
| **TR** Translation contact translation | 2 | TR (2, 0, 1) | Object sandwiched between wall |
| **OT** One-way translation contact translation | 1.5 | OT1 (1, 1, 1)  OT2 (0, 2, 1) | Object sandwiched between three walls |
| **PR** Prismatic contact translation | 1 | PR (1, 0, 2) | Drawer pulled out halfway |
| **OP** One-way prismatic contact translation | 0.5 | OP (0, 1, 2) | Drawer pushed all the way to the bottom |
| **FT** Fully contact translation | 0 | FT (0, 0, 3) | Totally immobile object |

| State name | DOFs | Admissible translation directions on the Gaussian sphere | Examples |
|---|---|---|---|
| **NR** Non-contact rotation | 3 | NR (3, 0, 0) | |
| **RT** Partial contact rotation | 2.5 | RT1 (2, 1, 0)  RT2 (1, 2, 0)  RTN (0, 3, 0) | |
| **SP** Spherical contact rotation | 2 | SP (2, 0, 1) | |
| **OS** One-way spherical contact rotation | 1.5 | OS1 (1, 1, 1)  OS2 (0, 2, 1)  OSN (0, 2, 1) | |
| **RV** Revolute contact rotation | 1 | RV (1, 0, 2) | Partially opened faucet (middle position) |
| **OR** One-way revolute contact rotation | 0.5 | OR (0, 1, 2) | Closed faucet |
| **FR** Fully constraint rotation | 0 | FR (0, 0, 3) | |

(a) Translation states                              (b) Rotation states

**Fig. 4.4** Object contact states

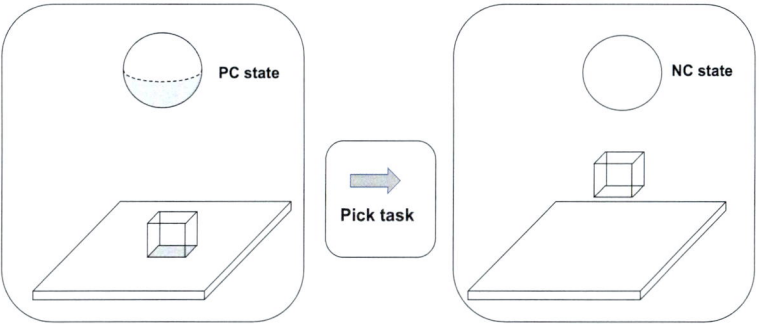

**Fig. 4.5** `Pick` task. The `Pick` task brakes the contact between the cube and the table and causes the `PC-NC` transition

is to cause the transition in the object's contact state from a hemispherical range of motion to one encompassing the entire sphere. In other words, the `PC-NC` transition is caused by the pick task. As illustrated by this example, we define a single task as a unit of actions that aims to generate one transition in the object's contact state.

Since the number of contact states is finite, the number of possible tasks, given by their transitions, is also finite. Given that there are seven states before the transition and seven states after the transition, there are 49 possible transitions and correspondingly 49 possible tasks.[2] This discussion applies to both translational and rotational motions, resulting in an upper bound of 98 tasks in the task set. Consequently, theoretically, we need to prepare 98 manipulation skills in the library to perform these possible tasks.

Physical analysis [9, 135] further reduces the maximum number of possible transitions. The previous 98 transitions encompass physically impossible transitions, such as an object

---

[2] In a unit of actions corresponding to a transition within the same state, it is assumed brief stops exit before and after the actions.

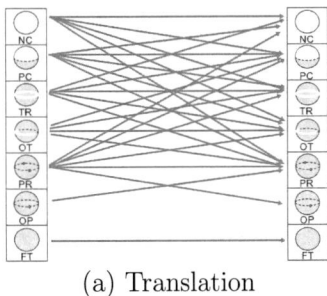
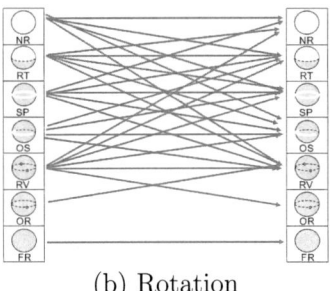

(a) Translation                                      (b) Rotation

**Fig. 4.6** State transitions and physically possible tasks

completely surrounded by walls on all sides (FC state) suddenly transitioning to an uncon-
fined, freely movable state (NC state). By evaluating the physical plausibility of each transition, we can determine that the feasible transitions correspond to the edges in the graph
shown in Fig. 4.6, which include 27 translational and 27 rotational transitions, making a total
of 54 transitions. Therefore, if we define tasks as a unit of actions that induce contact-state
transitions, the maximum number of tasks is 54. Consequently, the number of skills required
to perform these tasks to be prepared in the skill library also amounts to 54.

Furthermore, we analyzed the practical occurrence of these 54 tasks in YouTube videos.
The results are summarized in Table 4.1, which indicates that most tasks occur infrequently.
This analysis concludes that the nine tasks illustrated in Fig. 4.7–Bring (NC-NC), Pick
(PC-NC), Place (NC-PC), Drawer-adjust (PR-PR), Drawer-open (PR-OP),
Drawer-close (OP-PR), Door-adjust (RV-RV), Door-open (OR-RV), and
Door-close (RV-OR)–are sufficient for covering daily activities.[3] Consequently, in
LfO2.0, we design nine skills that perform these nine tasks in the current manipulation
skill library.

In the aforementioned discussion, the scope of the YouTube analysis was limited to
common household tasks, which restricted the observed tasks. In industrial assembly, tasks
such as inserting a rod into a hole (NC-PR) or completely extracting an inserted rod (PR-NC)
are common. Additionally, tasks such as inserting a new book between books on a bookshelf
(NC-TR) or removing a book from a bookshelf (TR-NC) occur frequently. However, these
tasks require visual feedback and skilled techniques, which might explain their infrequent
occurrence in household tasks. If necessary, any required skills from the remaining 48 types
not included in this library can be added.

---

[3] The PC-PC (Wipe) task will be explained in the next section.

**Table 4.1** Physically possible tasks and their frequency of occurrence

| Translation | | | | Rotation | | | |
|---|---|---|---|---|---|---|---|
| NC–NC | Bring | 21% | | OR–RV | Door-open | 6% |
| PC–NC | Pick | 23% | | RV–OR | Door-close | 6% |
| NC–PC | Place | 23% | | RV–RV | Door-adjust | 0% |
| PC–PC | Wipe | 20% | | NR–NR | | 0% |
| OP–PR | Drawer-close | 6% | | RT–NR | | 0% |
| PR–OP | Drawer-open | 6% | | NR–RT | | 0% |
| PR–PR | Drawer-adjust | 0% | | RT–RT | | 0% |
| OT–TR | | 0% | | OS–SP | | 0% |
| TR–OT | | 0% | | SP–OS | | 0% |
| PC–TR | | 0% | | RT–SP | | 0% |
| TR–NC | | 0% | | SP–NR | | 0% |
| TR–PC | | 0% | | SP–RT | | 0% |
| TR–TR | | 0% | | SP–SP | | 0% |
| OT–OT | | 0% | | OS–OS | | 0% |
| OT–PC | | 0% | | OS–RT | | 0% |
| OT–PR | | 0% | | OS–RV | | 0% |
| PR–OT | | 0% | | RV–OS | | 0% |
| PR–NC | | 0% | | RV–NR | | 0% |
| PR–PC | | 0% | | RV–RT | | 0% |
| PR–TR | | 0% | | RV–SP | | 0% |
| OT–NC | | 0% | | OS–NR | | 0% |
| NC–TR | | 0% | | NR–SP | | 0% |
| PC–OT | | 0% | | RT–OS | | 0% |
| PC–PR | | 0% | | RT–RV | | 0% |
| TR–PR | | 0% | | SP–RV | | 0% |
| NC–OT | | 0% | | NR–OS | | 0% |
| NC–PR | | 0% | | NR–RV | | 0% |

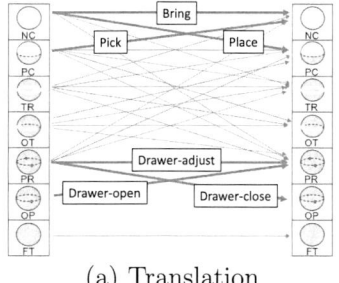

(a) Translation

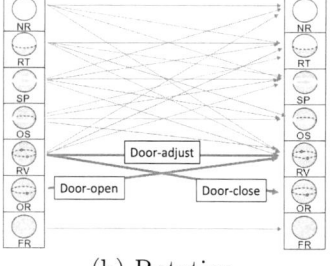

(b) Rotation

**Fig. 4.7** Physical manipulation tasks

## 4.2  Semantic Manipulation

Some household tasks require a unit of actions that incorporate additional constraints derived from common sense. These tasks cannot be defined solely by the physical constraints described in the previous subsection. For example, when wiping a table surface with a sponge, common sense dictates that the sponge should not leave the surface to clean it effectively. Physically, the sponge is merely in contact with the table's surface. In terms of physical constraints, it is in a `PC-PC` (point contact to point contact) transition state. In fact, the tasks observed as `PC-PC` in Table 4.1 correspond to this scenario.

To express the common sense constraint of "not leaving the surface," we introduce a virtual surface above the sponge and define the sponge's movement constraints as both the physical constraint surface and this virtual constraint surface. The sponge's movement is then expressed as a `TR-TR` (two-surface contact) task, moving only between these two surfaces. We refer to this constraint from the virtual surface as a semantic constraint, and tasks constrained by this as semantic tasks.

Figure 4.8 illustrates five types of semantic constraints identified from YouTube videos. The first type, the semantic pin, is observed in actions such as carrying a cup filled with juice. Physically, when transporting an object to a different location, its intermediate posture can freely rotate. However, a cup filled with juice can rotate around a virtual pin, but rotating around other axes would result in spillage. To represent this, we introduced the concept of a virtual pin inserted into the object, allowing rotation only around this axis. For translation, three degrees of freedom (DOF) are permitted, while rotation is restricted to one DOF, resulting in controlled translational movement.

The semantic walls allow two DOFs in translational motion, as previously discussed. In this case, rotation is also restricted to one DOF.

Translational motions with one DOF can be represented as the semantic tube. In scenarios such as cooking, where a peeling motion requires fixed directional movement, this motion can be represented as a tunnel-like constraint, corresponding to a semantic tube, where both vertical and horizontal movements are restricted. In this case, there is no rotational freedom.

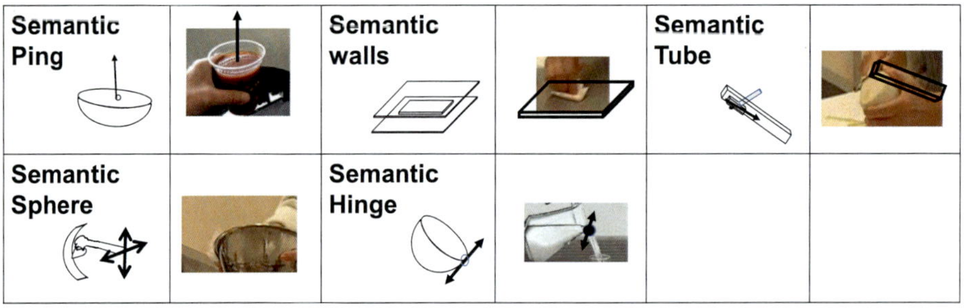

| Semantic Ping | | Semantic walls | | Semantic Tube | |
|---|---|---|---|---|---|
| Semantic Sphere | | Semantic Hinge | | | |

**Fig. 4.8**  Semantic constraints

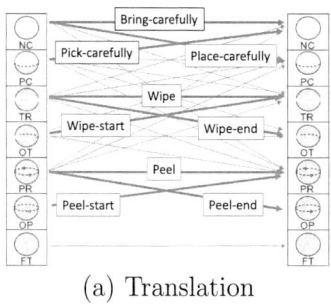
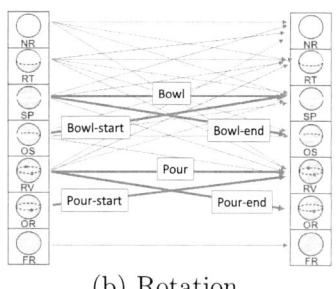

(a) Translation                                    (b) Rotation

**Fig. 4.9** Semantic manipulation tasks

For two DOFs in rotational motion, we have the semantic sphere, applicable to tasks such as wiping the inside of a spherical surface. The translational motion of the rotation center is fixed.

Lastly, there is the semantic hinge, relevant for one DOF rotational motion. For example, when pouring water from a pitcher into a cup, assuming an axis in space, rotating the pitcher around this axis to pour water into the cup represents this motion. In this case, the translational motion of the rotation center is also fixed.

From these semantic constraints, five semantic skill groups, as illustrated in Fig. 4.9, are obtained. The skill group comprising `Bring carefully`, `Pick carefully`, and `Place carefully` is derived from the semantic pin. The semantic plane and semantic tube provide the `Wipe` and `Peel` skill groups, respectively. The semantic sphere and semantic hinge derive the `Bowl` and `Pour` skill groups, respectively.

In the current semantic skill library, the frequently utilized skill groups, the `Bring-carefully` group, the `Wipe` group, and the `Pour` group, have been implemented.

## 4.3    Reward Function and Its Design

Manipulation skill agents–whether for physical or semantic tasks–are trained using reinforcement learning, similar to grasp skill agents. At this stage, instead of designing the reward function in an ad hoc manner, the relevant directions should be systematically examined based on the transition of contact states in each direction. We consider, each feasible motion direction under the object's constraint state. The reward function is then designed according to the state transitions that occur along the motion direction and its perpendicular directions.

To design a reward function based on the object's contact state along a motion direction and its perpendicular direction, each direction is described according to how the contact state changes as the object moves in that direction. To distinguish this from the general contact state of the object, the overall contact state is referred to as the "object contact state," while the state associated with a specific direction is referred to as the "directional contact state."

## 4.3.1  Directional Contact State

To examine the relationship between the contact state of an object and the reward function, each direction is designated according to the manner in which the object's contact state transitions   due to infinitesimal movement in that direction.

- **Maintenance direction**: A direction in which infinitesimal motion preserves the current contact state.
- **Detachment direction**: A direction in which infinitesimal motion results in the disappearance or appearance of the current contact.
- **Constraint direction**: A direction in which motion is restricted due to the existing contacts in that direction.

Each contact state, as depicted in Fig. 4.10a, can be represented using the DOF in the subspaces spanned by these directions. In a three-dimensional space, there are three DOF for both translation and rotation, leading to a total of three. As illustrated in Fig. 4.10a, the DOF values of each dimension, which sum to three, are categorized in dictionary format.

As an illustrative example of this subspace analysis, consider the object placed on the table, depicted to the right in Fig. 4.10b, whose contact state is PC1. When this object moves along the table's surface, there is no alteration in the contact state in the direction of motion. In other words, the object exhibits two DOF in the maintenance dimension, a subspace spanned by maintenance directions. Movement in a direction that incorporates the table's normal vector, specifically upward motion, disrupts the surface contact. The pure detachment direction, excluding the components of the surface contact direction, aligns with the table's normal direction. Therefore, the DOF in the detachment dimension is one. Since there is no direction in which the motion is constrained, the DOF in the constraint dimension is zero. Thus, the state of this object can be described using the number of maintenance, detachment, and constraint dimensions as (M=2, D=1, C=0).

Similarly, consider the case where an object is sandwiched between two planes (TR), as shown in Fig. 4.10c. In this scenario, motion along the plane is possible in two dimensions between the two planes, resulting in two DOF in the maintenance dimension. Conversely, motion perpendicular to the planes is restricted by the planes, resulting in one DOF in the constraint dimension. There are no detachment directions, i.e., zero DOF in the detachment dimension. Therefore, the contact state of this object can be described as (M=2, D=0, C=1).

| Translation | NC | PC1 | PC2 | PCN | TR | OT1 | OT2 | PR | OP | FT |
|---|---|---|---|---|---|---|---|---|---|---|
| Rotation | NR | RT1 | RT2 | RTN | SP | OS1 | OS2 | RV | OR | FR |
| Diagram on Gaussian sphere | ◯ | ◯ | ◯ | ◯ | ◯ | ◯ | ◯ | ◯ | ◯ | ◯ |
| Maintenance | 3 | 2 | 1 | 0 | 2 | 1 | 0 | 1 | 0 | 0 |
| Detachment | 0 | 1 | 2 | 3 | 0 | 1 | 2 | 0 | 1 | 0 |
| Constraint | 0 | 0 | 0 | 0 | 1 | 1 | 1 | 2 | 2 | 3 |

(a) Object contact states and their dimensions.

(b) PC1 contact state.

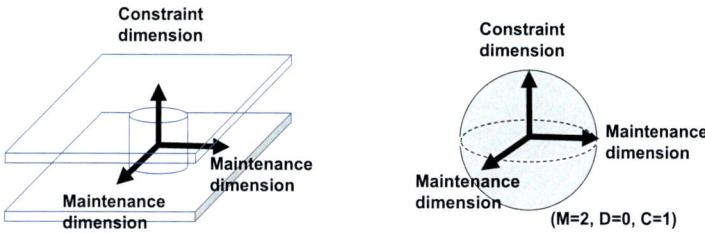

(c) TR contact state.

**Fig. 4.10** Object contact-state and dimension contact-state

## 4.3.2   Directional Transition and Control Policy

By characterizing tasks using directional transitions, we can derive the control rules necessary to accomplish these tasks. We define a task as a unit of actions that induces transitions in the contact state of objects. Within the framework of object contact transitions, it is possible to examine the dimensions of the transitions that occurred, leading to the object contact transition. This directional transition provides the conditions required to achieve the task.

For instance, consider the `Place` task, which involves positioning a box from the air onto a table along the vertical axis. In this scenario, the object's contact state before execution in the air is non-contact (NC), represented as (M=3, D=0, C=0). Post execution, the object rests on the table, with its contact state now being partial contact (PC), represented as (M=2, D=1, C=0). During this transition, one of the three DOF in the maintenance dimension, specifically the DOF along the vertical direction (the motion direction), is lost, and one DOF in the detachment dimension appears in that direction. In other words, the Place task is defined as a unit of actions that transitions one DOF in the maintenance dimension along the motion direction to one DOF in the detachment dimension.

Control policies for a skill can be derived from directional state transitions. In the previous example, there is no reactive force from the environment in the maintenance state along the motion direction. In contrast, in the detachment state, a reactive force is present in this motion direction. This indicates that by observing the reactive force from the environment along the motion direction, the completion of the Place task execution can be detected.

Similarly, consider the case where an object is constrained between two planes, as shown in Fig. 4.10c. In this scenario, motion is only possible in the direction of the maintenance dimension. When moving along a maintenance direction, the directional state remains in maintenance, and no reactive force from the environment occurs. Conversely, the directional state in the direction perpendicular to the motion remains constrained. Thus, even a small component of motion in this direction will cause a reactive force from the constraining plane. Therefore, to successfully move the object along the maintenance dimension, a unit of actions must be designed to ensure that no reactive force occurs along the perpendicular direction due to small disturbances.

In the following discussion, the motion direction is denoted as S, while the directions perpendicular to the motion are denoted as T and U. Additionally, in future discussions, it may be necessary to evaluate whether a transition has occurred or not. For this purpose, we introduce the flag BeforeTransition, which is True before the transition and False after the transition.

### 4.3.3  Reward in the Motion Direction

This subsection will examine the control policies in the motion direction. As shown in Fig. 4.11, four transitions occur in the motion direction. Three real transitions can be considered: maintenance-maintenance, maintenance-detachment, and detachment-maintenance. If the state is constrained in the motion direction, no motion can occur in that direction, and thus no transitions to or from this state are possible. However, one virtual transition, constraint-constraint, is also included for the sake of completeness.

We will define control policies for these four dimension transitions sequentially. To be compatible with reinforcement learning, we will use notation similar to reward functions in reinforcement learning. If reinforcement learning is not required, the condition expressions themselves will serve as termination conditions.

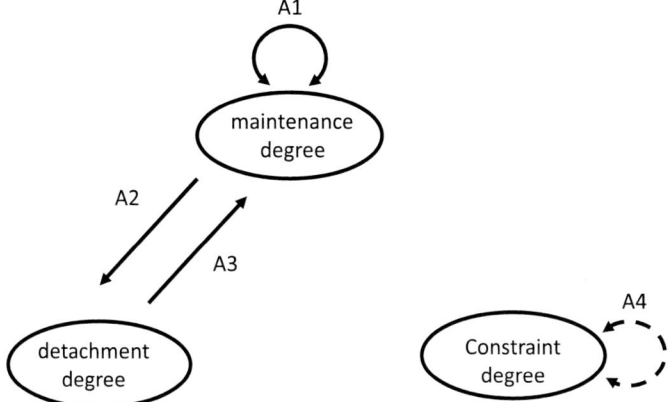

**Fig. 4.11** Dimensional state transitions in the motion direction

**A1: Maintenance-Maintenance** In the motion direction, the directional state remains in the maintenance state, with no reactive force from the environment occurring throughout the entire movement. Therefore, position control can be utilized, and the termination condition can be defined using the s-coordinate value of the goal position, `goal_s`, given from the demonstration.

```
A1: if s = goal_s, then reward
```

**A2: Maintenance-Detachment** Along the motion direction, an object in the maintenance directional state experiences no reactive force from the environment. When the directional state transitions to detachment, a reactive force arises in that direction. This implies that a collision with a surface occurs during the motion.

For a control policy, even if the goal position is specified from the demonstration, considering the possibility of observation errors, it is more robust to use the position at which the reactive force is detected as the termination position. Given the threshold `delta-zero` for the force sensor reading along the s coordinate system, `F_s`, the termination condition can be described as follows:

```
A2: if F_s > delta-zero, then reward
```

**A3: Detachment-Maintenance** The reverse of A2 occurs when an object moves in the direction of the detachment dimension. This movement causes the surface contact to disappear, transitioning the object to the maintenance directional state. To confirm the transition to the maintenance state, it may be valid to include the condition that the reactive force becomes zero. Although this transition involves only an infinitesimal movement, there is a finite movement interval in typical robotic control during which the object moves in the maintenance state. Therefore, considering that the finite interval is long enough to avoid the influence of observational errors, the termination condition can be based solely on the position given by the demonstration.

```
A3: if s = goal_s, then reward
```

**A4: Constraint-Constraint** If the motion direction is in the constraint dimension, movement cannot occur; thus this transition does not exist physically. However, it can be used in cases of semantic constraints, so it is defined here for the sake of completeness.

```
A4: if F_s > delta-collision, penalty
```

### 4.3.4   Reward in the Orthogonal Direction

Regarding the directional state perpendicular to the motion direction, transitions are possible between all directional states, and nine transitions can be considered, as shown in Fig. 4.12. In the following discussion, although transitions occur through infinitesimal movements, we will consider finite movement intervals before and after the transitions, taking into account the practical operation of a robot.

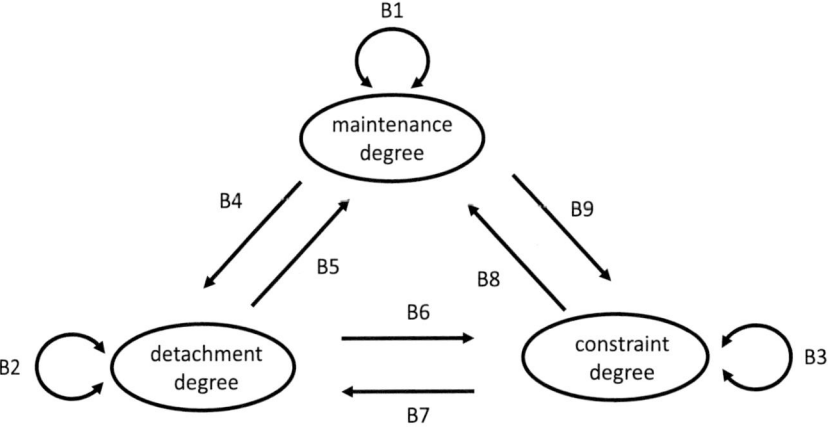

**Fig. 4.12** Dimensional state transitions in the orthogonal direction to the motion

**B1: Maintenance-Maintenance** For the direction T that is perpendicular to the motion direction, when the direction is in the maintenance state, the object is not constrained by the environment in the T direction. Therefore, positional errors in that direction are permissible, and the termination position of the skill in the T direction is determined solely by the demonstration. Here, `goal-t` represents the termination position in the T dimension given by the demonstration.

```
B1: if T = goal_t, then reward
```

**B2: Detachment-Detachment** For the perpendicular direction T to the motion direction, maintaining the detachment state is equivalent to maintaining surface contact in the T direction during the motion. Therefore, it is necessary to keep the reactive force from the environment within a certain range during the motion. In other words, the movement direction must be adjusted to prevent the reactive force from becoming too large, leading to a collision, or too small, causing the object to detach from the surface.

```
B2: if F_t > delta-collision, then penalty
    if F_t < delta-zero, then penalty
```

**B3: Constraint-Constraint** To maintain the constraint state in the orthogonal direction T, it is necessary to adjust the motion direction so that the reactive force from the environment in the direction T is minimized, similar to B2. However, since there is no detachment from the constrained state in the orthogonal direction, the second condition equivalent to the `delta-zero` condition in the case of B2, to keep the contact, is not necessary. The termination condition is provided by other dimensions.

```
B3: if F_t > delta-collision, then penalty
```

**B4: Maintenance-Detachment** When maintaining the state, there is no contact and the direction is airborne. However, after the transition, the detachment state implies surface contact occurs. Visual sensors are utilized to adjust the motion direction to achieve surface contact after the transition. The position coordinates along this direction of the contact surface, referred to as `feature-t`, such as the coordinates of the boundary edge of the environmental object, need to be aligned with the position T of this object in this direction using visual information. After the transition, the motion direction is adjusted to maintain the detachment state using conditions similar to B2 within a finite interval. The termination condition is provided by other directions.

```
B4: if BeforeTransition:
        if |T - feature_t| > delta-gap, then penalty
    else:
        if F_t > delta-collision, then penalty
        if F_t < delta-zero, then penalty
```

**B5: Detachment-Maintenance** In the finite interval of the detachment state before the transition, the motion direction is adjusted to maintain surface contact using B2 to ensure that the reactive force from the perpendicular direction, T remains within a certain range during motion. In the maintenance state after the transition, there are no conditions imposed on this dimension, allowing for free positioning. Assuming this finite interval is sufficiently long, the positional errors between the execution and demonstration times in this dimension can be corrected by the termination condition. Therefore, similar to B1, the position information goal-T provided during the demonstration is used as the termination condition for this dimension.

```
B5: if BeforeTransition:
        if F_t > delta-collision, then penalty
        if F_t < delta-zero, then penalty
    else:
        if T = goal_t, then reward
```

**B6: Detachment-Constraint** A transition from the detachment state to the constraint state occurs at some point, even without special control measures. If this transition does not happen, there is an issue with the external shape, making the command un-executable. After the transition to the constraint state, control policies specific to the constraint state may be applied; however, detecting the transition requires considerable effort. As maintaining control policy for the detachment state is sufficient to uphold the constraint state, control policies for the detachment state is applied throughout the entire interval. The termination condition is determined by other dimensions.

```
B6: if F_t > delta-collision, then penalty
    if F_t < delta-zero, then penalty
```

**B7: Constraint-Detachment** This process is the inverse of B6. Generally, maintaining the constraint state necessitates controlling to minimize the lateral reactive force. Post-transition, the state transitions to the detachment, requiring directional control to ensure the lateral reactive force remains below a specific threshold, yet not zero. To mitigate the cost associated with transition detection, similar to the approach in B6, control policies for the detachment state are maintained throughout the entire interval. The termination condition is determined by other dimensions.

```
B7: if F_t > delta-collision, then penalty
    if F_t < delta-zero, then penalty
```

**B8: Constraint-Maintenance** In the constraint state, the control policy described in B3 is applied. Following the transition from the constraint state to the maintenance state, positional control becomes feasible. Hence, similar to B1, the terminal condition is defined by positional information, given by the demonstration.

```
B8: if BeforeTransition:
        if F_t > delta-collision, then penalty
    else:
        if T = goal-t, then reward
```

**B9: Maintenance-Constraint** In the maintenance state, there is no force information, and the motion direction is adjusted using visual information to transition to the constraint state. After the transition, the constraint state is maintained with the control policies similar to B3.

```
B9: if BeforeTranstion:
        if |T - feature_t| > delta-gap, then penalty
    else:
        if F_t > delta-collision, then penalty
```

## 4.4    Implementing Reward Functions (Physical Skills)

This section obtains actual reward functions for reinforcement learning of each skill agent based on the policies described in the previous section, assuming all constraints are actual physical constraints.

Based on the transitions of directional states, we design a reward function for reinforcement learning to achieve each task by combining partial policies from the previous section. In this process, we use the OR operation for penalty terms and the AND operation for reward terms. This is because even a single unmet penalty can hinder the operation, while achieving the goal requires meeting all reward conditions.

### 4.4.1    Reward Functions for `Pick/Bring/Place` group

This group represents the most fundamental manipulation task for robots and occurs with the highest frequency. Differences among the three tasks in this group arise concerning the transition of directional states in the motion direction. All transitions in directional states that are orthogonal to the motion are common across these tasks.

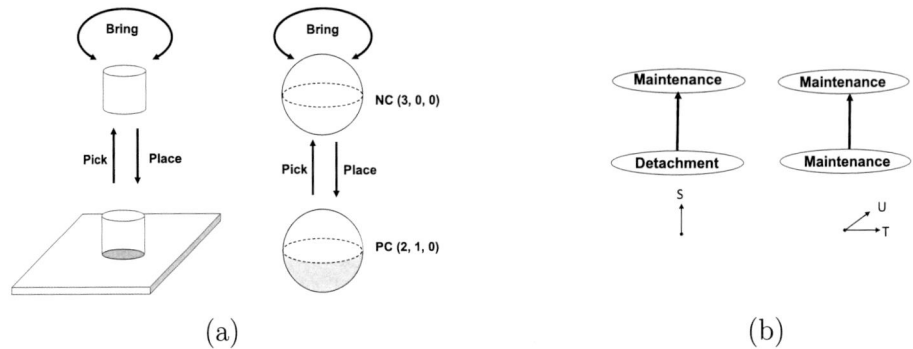

**Fig. 4.13** Pick/Bring/Place tasks. **a** Object state transitions in Pick/Bring/Place tasks. **b** Directional state transitions in Pick task

**Reward function for Pick skill agent** The Pick task is defined as a unit of actions that causes a transition in the contact state of an object from (M=2,D=1,C=0) to (M=3,D=0,C=0). See Fig. 4.13a.

During this transition, in the motion direction S, the directional contact state transitions from the detachment to the maintenance, while in the orthogonal directions T and U, both directional states remain in the maintenance. See Fig. 4.13b.

In the motion direction, A3 can be applied, and the reward function can be defined using the terminal position given by the demonstration. As stated in A3, when the Pick is executed, the reactive force along the movement disappears, but it is assumed that the target position is sufficiently distant from the observation error range, and the position provided by the demonstration is used solely as the target position.

```
A3: S = goal_s, then reward
```

In the orthogonal directions, T and U, the maintenance state persists, thus allowing for the application of B1. The termination conditions for these dimensions are defined using the demonstrated position in these directions.

```
B1: if T = goal-t, then reward
B1: if U = goal_u, then reward
```

Since all the terms are reward components, we combine them using the AND operation to obtain the following function.

```
Reward function for Pick skill agent
    if S = goal_s AND T = goal_t AND U = goal_u, then reward
```

**Reward function for `Bring` skill agent** The same control policies can be applied to the Bring skill agent as well:

```
Reward function for Bring skill agent
    if S = goal_s AND T = goal_t AND U = goal_u, then reward
```

**Reward function for `Place` skill agent** The `Place` task can be defined as a unit of actions that causes the transition in the object contact state from (M=3, D=0, C=0) to (M=2, D=1, C=0).

In the motion direction S, a transition from the maintenance to the detachment occurs. Consequently, A2 is applicable. It is important to note that the terminal position derived from the demonstration may contain errors. Therefore, employing force feedback to determine the stop position is preferable to avoid issues arising from collisions.

```
A2: if F_s > delta-zero, then reward
```

Regarding the orthogonal directions to the motion, T and U, the maintenance state persists, thus allowing for the application of B1. Namely, the position information obtained from the demonstration is employed to determine the stop position.

```
B1: if T = goal_t, then reward
B1: if U = goal_u, then reward
```

By combining all reward components using the AND operation, we obtain:

```
Reward function for Place skill agent
    if F_s > delta-zero AND T = goal_t AND U = goal_u, then reward
```

## 4.4.2   Reward Functions for `DrawerOpen/Adjust/Close` group

These three tasks are commonly classified under the compliance motion group, characterized by constraints in directions orthogonal to the motion. Differences exist in the state transition within the motion direction, similar to those observed in the previous `Pick/Bring/Place` group.

**Reward function for `Drawer-open` skill agent** During the Drawer-open task, the object contact state transitions from (M=0, D=1, C=2) to (M=1, D=0, C=2). See Fig. 4.14.

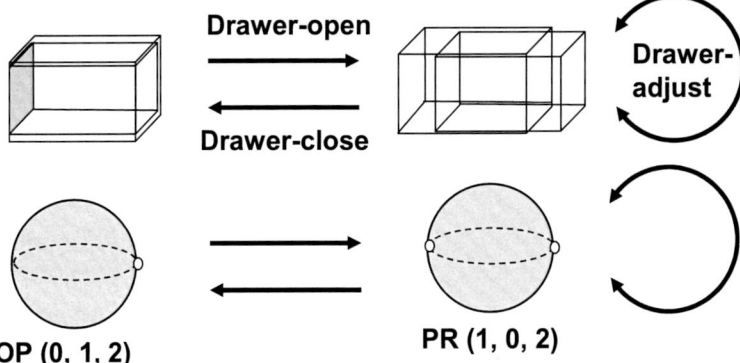

OP (0, 1, 2)                          PR (1, 0, 2)

**Fig. 4.14** Drawer-open/adjust/close tasks

Regarding the motion direction S, a transition from the detachment to the maintenance occurs. Therefore, A3 can be applied, and the agent uses the position obtained from the demonstration to determine the stop position.

```
A3: if S = goal_s,
        then reward
```

The orthogonal directions T and U to the motion, they remain constrained. According to B3, the agent needs to adjust the direction of movement to minimize the reactive force in these directions as much as possible.

```
B3: if F_t > delta-collision, then penalty
B3: if F-u > delta-collision, then penalty
```

By parallelizing the penalty terms using the OR operation and separating them from the reward term, the reward function for the agent is given as:

```
Reward function for Drawer-open skill agent
    if F_t > delta-collision, penalty
    if F-u > delta-collision, penalty
    if S = goal_s, then reward
```

**Reward function for DrawerAdjust skill agent** During the drawer-adjust task, the object's contact state remains (M=1, D=0, C=2). The directional state along the motion direction remains the same, A1 can be applied, and the agent can use the ending position from the demonstration.

```
A1: S = goal_s, then reward
```

In the perpendicular directions T and U, the constraint state continues. B3 can be applied, and in these directions, the agent needs to adjust the motion direction to minimize the resistance force as much as possible.

```
B3: if F_t > delta-collision, then penalty
B3: if F_u > delta-collision, then penalty
```

Combining these partial policies, the reward function for the agent is given as:

```
Reward for DrawerAdjust skill agent
    if F_t > delta-collision, then penalty
    if F_u > delta-collision, then penalty
    if S = goal_s, then reward
```

**Reward function for DrawerClose skill agent**  During the DrawerClose task, the object's contact state transitions from (M=1,D=0,C=2) to (M=0,D=1,C=2). The directional state along the motion direction transitions from maintenance to detachment. A2 can be applied, and similarly to the Place task case, the reward function utilizes the occurrence of the resistance force in this dimension as the stop position.

```
A2: if F_s > delta-zero, then reward
```

In the orthogonal directions T and U, the reward function is designed to minimize the resistance force as much as possible, in accordance with the B3 rule.

```
B3: if F_t > delta-collision, then penalty
B3: if F_u > delta-collision, then penalty
```

Combining these,

```
Reward for DrawerClose skill agent
    if F_t > delta-collision, penalty
    if F_u > delta-collision, penalty
    if F_s > delta-zero, reward
```

### 4.4.3   Reward Functions for DoorOpen/Adjust/Close group

This group exhibits a common characteristic in terms of compliance motion relative to the direction of the rotation axis. The transitions in the state of rotational direction categorize this group into three distinct cases: Door-open, Door-adjust, and Door-close.

By reinterpreting the rotational direction as the motion direction and the rotation axis direction as the orthogonal directions to the motion, a similar discussion can be developed

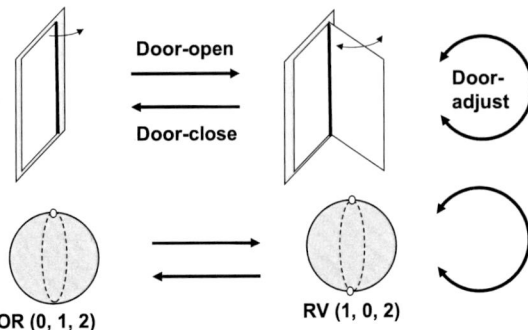

**Fig. 4.15** DoorOpen/Adjust/Close tasks

as with the Drawer group. For the discussion of infinitesimal rotation, the coordinate system A, B, C is used. These correspond to the coordinate system, S, T, U for infinitesimal translation. For further details, refer to [135].

**Reward function for DoorOpen skill agent** The contact state of the object with respect to rotational motion transitions from (M=0, D=1, C=2) to (M=1, D=0, C=2). See Fig. 4.15.

For the direction A along the opening motion, the reward function involves position control based on the demonstration position due to A1. For the orthogonal directions B and C relative to the motion direction, the reward function is based on force information to minimize resistance as much as possible due to B3.

```
Reward function for Door-open skill agent
    if F_b > delta-collision, penalty
    if F_c > delta-collision, penalty
    if A = goal_a, reward
```

**Reward function for DoorAdjust skill agent** For the DoorAdjust task, the same principles as the DrawerAdjust task can be applied. In the motion direction A, the directional state remains in the maintenance, and A1 can be applied, using the ending position to formulate the reward function. In the orthogonal directions B and C, the constraint state continues. Therefore, based on B3, the reward function is designed to minimize constraints as much as possible with respect to force information in these dimensions.

```
Reward function for DoorAdjust skill agent
    if F_b > delta-collision, penalty
    if F_c > delta-collision, penalty
    if A = goal_a, reward
```

**Reward function for** `DoorClose` **skill agent** For the `DoorClose` task, similar to the `DrawerClose` task, the object's contact state transitions from (M=1,D=0,C=2) to (M=0,D=1,C=2). In the motion direction A, the state transitions from maintenance to detachment, and A3 can be applied. Similar to the Place case, the reward function includes the generation of the resistance force in this direction. On the other hand, in the orthogonal directions B and C to the motion direction, the reward function is designed to reduce the resistance force below a certain threshold.

```
Reward function for DoorClose skill agent
    if F_b > delta-collision, penalty
    if F_c > delta-collision, penalty
    if F_a > delta-zero, reward
```

## 4.5   Implementing Reward Functions (Semantic Skills)

In some semantic tasks, translational motion is the primary focus, with the condition that rotational motion is constrained, and vice versa. Consequently, it is necessary to consider control policies for both translational and rotational motions simultaneously. Furthermore, semantic constraints differ from physical constraints in that they are virtual, necessitating the introduction of virtual forces to account for other characteristics such as positional information.

### 4.5.1   Reward Functions for `PickCarefully`/`BringCarefully`/`PlaceCarefully` group

For this group, although translational motion is the primary focus, the semantic constraints arising from rotational motion also play a significant role. Regarding translational motion, we can apply the same policies as the `Pick`/`Bring`/`Place` group.

`Reward Function for PickCarefully skill agent` Regarding rotational motion, the object's contact state is semantically represented as (M=1,D=0,C=2). For the rotation direction, S, the directional state remains in maintenance, allowing A3 to be applied, with the terminal rotation angle, given from the demonstration, being part of the reward function. For the orthogonal directions, T and U to the rotation direction, the directional state remains constrained, allowing B3 to be applied. Since no physical resistance arises from the environment, a pseudo-resistance, such as F_t and F_u, is generated based on deviation in this direction, forming the reward function.

Regarding translational motion, as was the case in the Pick task, the state along the motion dimension transitions from detachment to maintenance, and the state in the dimensions perpendicular to the motion remains in maintenance. Therefore, A3 and B1 can be applied similarly to the Pick task case, using position information to design the reward function.

```
Reward function for pick-carefully skill agent
    if F_b > delta-collision, penalty
    if F_c > delta-collision, penalty
    if S = goal_s AND T = goal_t AND U = goal_u AND A = goal_a, reward
```

**Reward function for BringCarefully skill agent** Regarding rotational motion, the same discussion as Pick-carefully can be applied. That is, for the rotational direction, A3 is applicable, and for the direction orthogonal to it, B3 is applicable to the pseudo-forces.

For translational motion, the same policies as Bring can be applied. Specifically, A1 is applicable for the motion direction, and B1 is applicable for the orthogonal direction. Therefore,

```
Reward function for BringCarefully skill agent
    if F_b > delta-collision, penalty
    if F_c > delta-collision, penalty
    if S = goal_s AND T = goal_t AND U = goal_u AND A = goal_a, reward
```

**Reward function for PlaceCarefully skill agent** Regarding rotational motion, the same policies as the first two cases apply, specifically A3 and B3. For translational motion, in the direction of motion, the state transitions from maintenance to detachment; thus, A2 is applicable. For the directions orthogonal to the motion, the state remains in maintenance; hence, B1 is applied.

```
Reward function for PlaceCarefully skill agent
    if F_b > delta-collision, penalty
    if F_c > delta-collision, penalty
    if F_s > delta-zero AND T = goal_t AND U = goal_u AND A = goal_a, reward
```

### 4.5.2  Reward Functions for Wipe group

This group consists of three tasks: WipeStart, WipeAdjust, and WipeEnd. Specifically, WipeStart refers to not reverting from the starting point of the Wipe, while WipeEnd implies not operating beyond the endpoint. These points may be considered semantic boundaries, with their positions determined based on information from the demonstration. However, for implementation simplicity, these tasks are treated the same as WipeAdjust. Thus, all three tasks are integrated into a single task, referred to as Wipe.

In the case of the Wipe task, translational motion is the main movement, and semantic constraints also occur in this translational motion. Rotational motion has no constraints and remains in maintenance.

Regarding translational motion, under physical constraints, the object contact state is represented as $(M=2, D=1, C=0)$, while under semantic constraints, the contact state is represented as $(M=2, D=0, C=1)$. Namely, the directional state in one direction orthogonal to the motion transitions from detachment to constraint.

In the direction orthogonal to both the motion and the constraint surface, referred to here as $T$, the directional state remains in constraint under semantic constraints. However, since actual resistance forces do not arise from the semantic constraint surface, contact with the physical constraint surface must be maintained to prevent collisions with this virtual surface. Therefore, the policy of maintaining a detachment state with respect to the actual physical surface, specifically B2, is employed.

In another orthogonal direction to the motion, $U$ direction, which is parallel to the physical constraint plane, it remains in maintenance, and therefore, B1 can be applied, with the terminal position determined from the values in the demonstration.

In the motion direction $S$, it remains in maintenance, and thus A1 can be applied, and the terminal position is given from the demonstration position.

Regarding rotation, in the direction of motion and in the directions orthogonal to it, the directional states remain in maintenance and constraint, respectively. This allows B1 and B3 to be applied, respectively.

```
Reward function for Wipe skill agent
    if F_t > delta-collision, penalty
    if F_t < delta-zero, penalty
    if F_b > delta-colllsion, penalty
    if F_c > delta-collision, penalty
    if S = goal_s AND U = goal_u AND A = goal_a, reward
```

### 4.5.3  Reward Functions for Pour group

This group consists of PourStart, PourAdjust, and PourEnd. However, similar to the case of Wipe, they are collectively discussed under the Pour task.

In the case of the Pour task, semantic constraints are imposed on rotational motion. The rotational state, under semantic constraints, is represented as $(M=1, D=0, C=2)$.

Semantically, in the direction orthogonal to the rotational direction B and C, the directional state remains in constraint, so B3 can be applied, and the reward function aims to minimize resistance. However, since there is no actual resistance from the environment, pseudo-resistance is generated based on the deviation in the direction of the rotational axis, and the reward function is designed accordingly.

Semantically, the state remains in maintenance for the rotational motion direction A; thus, B1 can be applied, and the reward function is based on rotational angle information provided from the demonstration.

In this task, translational motion does not and must not occur. Consequently, the semantic constraint state is maintained in all directions S, T, and U, and A4 is applied based on the pseudo-resistance.

```
Reward function for Pour skill agent
    if k |b - goal_b| > delta-collision, penalty
    if k |c - goal_c| > delta-collision, penalty
    if k |s - goal_s| > delta-collision, penalty
    if k |t - goal_t| > delta-collision, penalty
    if k |u - goal_u| > delta-collision, penalty
    if a = goal-a, reward
```

## 4.6    Training Manipulation Skill Agents

Manipulation skill agents undergo pre-training to develop policies for adjusting motion directions based on a reward function. Figure 4.16 illustrates the execution phase of a pre-trained agent, showcasing its utilization of force feedback and skill parameters, in accordance with its policy.

For pre-training, we utilize a reinforcement learning environment that leverages OpenAI's Baseline PPO algorithm [59] and employs PyBullet [58] as the environment simulator. During the learning process, we assume that the object to be manipulated is already grasped

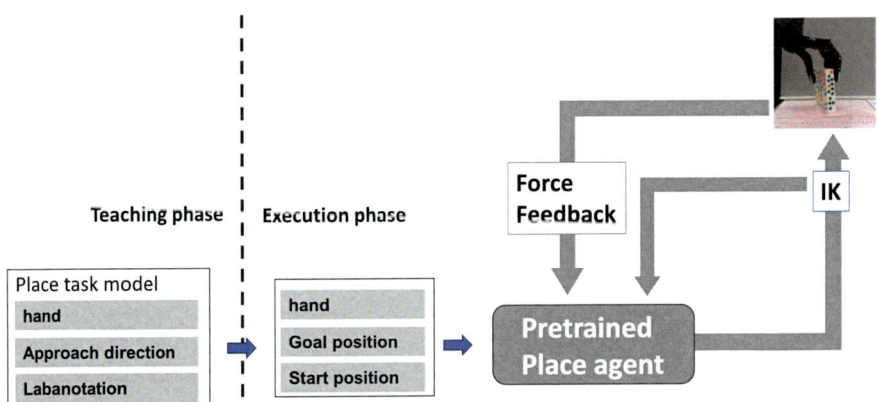

**Fig. 4.16** Manipulation skill agent. The agent determines the initial movement direction based on the skill parameter values. Hand motion is achieved through inverse kinematics (IK). The agent iterates a loop in which force feedback and joint angle information from the action are used as state variables to determine the subsequent hand movement according to the pre-trained policy

in the hand, integrating the hand and object. Policies are developed to adjust the direction of motion based on reaction forces. To account for the hardware's elasticity, contact is modeled as an elastic body, with spring and damper coefficients set accordingly. Additionally, to simplify the learning environment, gravity is set to zero. Consequently, the force sensor readings provided to the agent in the implementation are with gravity subtracted.

It is presumed that the actual robot is equipped with a 6-axis force sensor at the end of its arm. Due to discrepancies in physical parameters between the simulator and the actual machine, accurately simulating the magnitude of the force presents significant challenges. Consequently, to effectively apply the policy derived from the simulation to the actual robot, we elected to utilize the force direction unit vector, $\mathbf{f}_n$, instead of the raw values from the force sensor, $\mathbf{f}$, as state variables during reinforcement learning, whenever feasible.

$$\mathbf{f}_n = \mathbf{f}/|\mathbf{f}|. \tag{4.9}$$

Additionally, the force direction is transformed and expressed within the world coordinate system by leveraging the pose of the object obtained from the simulator, in conjunction with the Forward Kinematics (FK) of the actual robot.

When threshold determination is indispensable, the force magnitude is coarsely discretized and employed as a state variable. The parameter $f_{step}$, a discretization constant, dictates the level of discretization. Different discretization values are utilized depending on the specific circumstances to mitigate the sim-to-real gap.

$$f_{desc} = \lfloor |\mathbf{f}|/f_{step} \rfloor. \tag{4.10}$$

where $\lfloor x \rfloor$ is the floor function that rounds down the input number $x$ to the nearest integer.


### 4.6.1   Implementing `Pick/Bring/Place` skill agents

**`Pick/Bring` skill agents** `Pick/Bring/Place` skill agents generate the most basic unit of actions. The `Pick` and `Bring` agents can be controlled solely with position information.

First, let us take the `Pick` skill as an example to illustrate how to use the reward function provided in the previous subsection for designing the actual skill. Although it is referred to as a reward function for consistency, this skill can be executed using deterministic position control until it reaches the goal position; it does not require any reinforcement learning process during the design stage.

The decoder passes the following parameters to the agent:

```
Pick {hand}{departure}{labanotation}
```

Here, `hand` refers to which hand is being used, `departure` is a directional vector indicating the distance and direction of the `Pick` task, and `labanotation` represents human posture during this task. In this discussion, the relative movement vector `departure` is relevant to the subsequent arguments.

The reward function of the `Pick` skill agent was given as:

```
if S = goal_s AND T = goal_t AND U = goal_u, then reward
```

To complete this reward function, the goal position is determined based on the `departure` vector. The initial hand position before this pick task is retrieved from the end position of the previous task, usually a grasp task, by the decoder and provided to this agent as the starting position. The vector `departure` represents both the departure direction from the starting position and the distance from the starting position. Thus, the goal position, i.e., the values of `goal_s`, `goal_t`, and `goal_u`, can be calculated from the starting position and this relative movement vector `departure`.

Based on this reward function, the `Pick` skill agent gradually approaches the goal position from the current position. At each iteration, a new hand position is calculated toward the goal position from the current position. Inverse kinematics (IK) is performed at each step, and the new joint angles of the entire robot are calculated from the next hand position. In this case, when solving the IK for the first time, the initial posture given by the Labanotation in `{labanotation}` is used as the initial solution for the IK, and thereafter, the posture of the previous step is used as the initial solution for the IK. The skill agent completes its tasks once it reaches the goal position.

The `Bring` skill agent is also implemented in a similar manner.

**Place skill agent** The operation of the `Place` skill agent is also almost the same as that of the Pick skill agent, except for the termination condition. The Place skill is given the following parameters:

```
Place {hand}{approach}{labanotation}
```

The final position obtained from the demonstration, i.e., `approach`, is likely to include observational errors. Therefore, it also moves along `approach`, but it terminates when the resistance force in the direction exceeds a threshold value, rather than reaching the final goal position, `goal_s`.

```
if F_s > delta-zero AND T = goal_t AND U = goal_u, then reward
```

The following presents the execution results using an actual robot to verify the correct functioning of the implemented `Place` skill agent. Figure 4.17a illustrates the transitions in force sensor values. The horizontal axis represents time steps, while the vertical axis indicates the force sensor values in the direction of motion. To eliminate the effect of gravity, the force at the start of the skill is maintained, and the difference from the current sensor value is calculated. Additionally, the sensor values in the robot's coordinate system are transformed into force information in the world coordinate system using forward kinematics (FK).

It can be observed that when the object makes contact with the surface, resistance is generated, resulting in an increase in force. `delta-zero` is set to 3[N], and contact is detected at time step 12, indicating the successful completion of this skill. Figure 4.17b demonstrates the execution of the `Place` skill agent using this reward function.

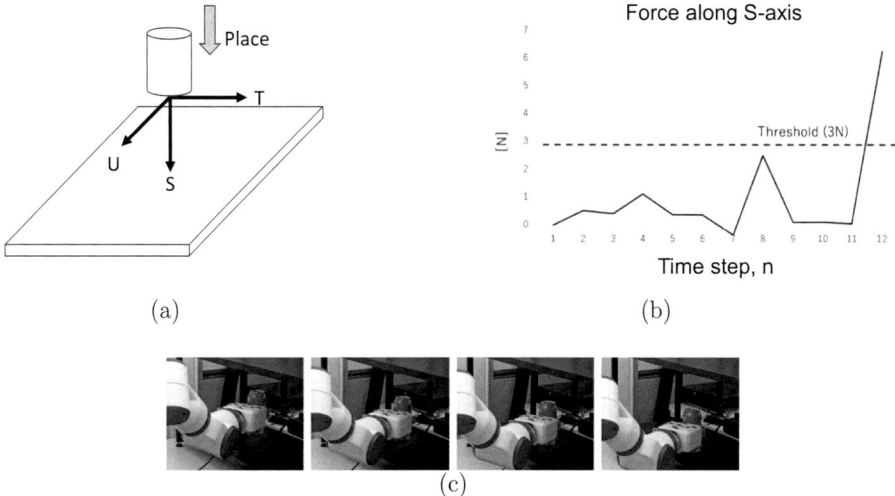

(a)                                                          (b)

(c)

**Fig. 4.17** Execution of pre-trained `Place` skill agent. **a** Coordinate system. **b** The magnitude of force along the S axis. **c** Execution of `Place` task by the agent

## 4.6.2   Implementing `DrawerOpen/Adjust/Close` skill agents

The `DrawerOpen/Adjust/Close` skill agents adjust the pulling/pushing direction based on the policy learned based on the reward function. The original pulling/pushing direction given as the skill parameter contains observation errors; it is necessary to adjust if a significant resistance force is generated in the direction perpendicular to the movement.

Ignoring the termination conditions, these agents can essentially follow a similar control policy. Therefore, to reduce the training effort, we first train the Drawer-adjust skill agent and then add a program to determine the termination conditions based on its policy, resulting in the Drawer-open and Drawer-close agents.

For the state values in reinforcement learning, the current motion vector $\mathbf{S}^n$ and the unit vector of the orthogonal resistance force $\frac{\mathbf{f}^n}{|\mathbf{f}^n|}$ to the motion direction at the time, $n$ are provided. $\mathbf{f}^n$, the observed resistance force is normalized for reducing the sim2real gap by avoiding the raw value.

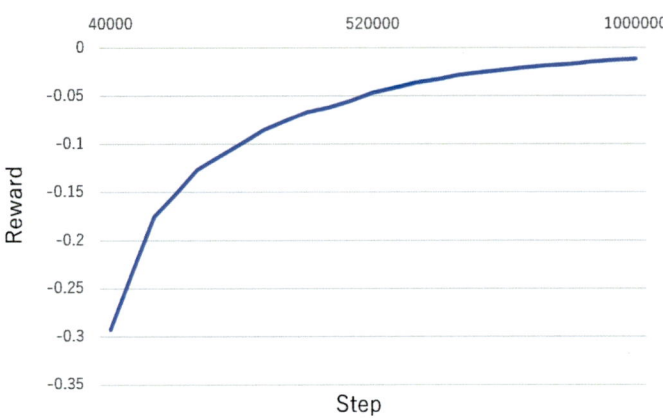

**Fig. 4.18** Learning curve of `DrawerAdjust` agent

The agent modifies the motion direction by $\Delta \mathbf{S}^n$. The motion direction at time $n+1$ is obtained as:

$$\mathbf{S}^{n+1} = \frac{\mathbf{S}^n + \Delta \mathbf{S}^n}{|\mathbf{S}^n + \Delta \mathbf{S}^n|}.$$

The penalty terms in the reward function,

```
if F_t > delta-collision, then penalty
if F_u > delta-collision, then penalty
```

are realized as the reduction of the reward as:

$$r = -|\mathbf{f}^n|. \tag{4.11}$$

It is important to note that raw values, $\mathbf{f}^n$ are used for training; however, these values are not employed during execution, thereby causing no issues related to the Sim2Real gap.

In reinforcement learning training, the episode ends after a predefined period has elapsed. Figure 4.18 shows the reward curve. The training concluded after 1 million steps.

For the termination conditions, the end position can be used for the `DrawerOpen` and `DrawerAdjust` agents. On the other hand, for the `DrawerClose` agent, the episode ends when the vertical resistance force in the pushing direction exceeds a certain threshold, `delta-zero`.

```
if F_s > delta-zero, reward
```

Figure 4.19 shows the execution of the pre-trained `DrawerClose` agent. Figure 4.19a depicts the force values in the orthogonal directions, $T$ and $U$, to the drawer's motion direction. Here, each displacement distance is set to 5 mm per step. During execution, unwanted resistance forces may be generated, but the agent adjusts the pushing direction to suppress force generation. This drawer has a locking mechanism that prevents it from opening automatically. Around the 22nd step in the right figure, instability is observed due to differences in physical conditions when approaching the locking mechanism. Figure 4.19b shows the force along the motion direction, S. A significant force is generated at step 21, just before the drawer closes, when the lock engages. This force then decreases, and a large force is generated again when the drawer is fully closed. Despite these locking mechanisms, as shown in Fig. 4.19c, the skill agent successfully executed the `DrawerClose` task on the real robot.

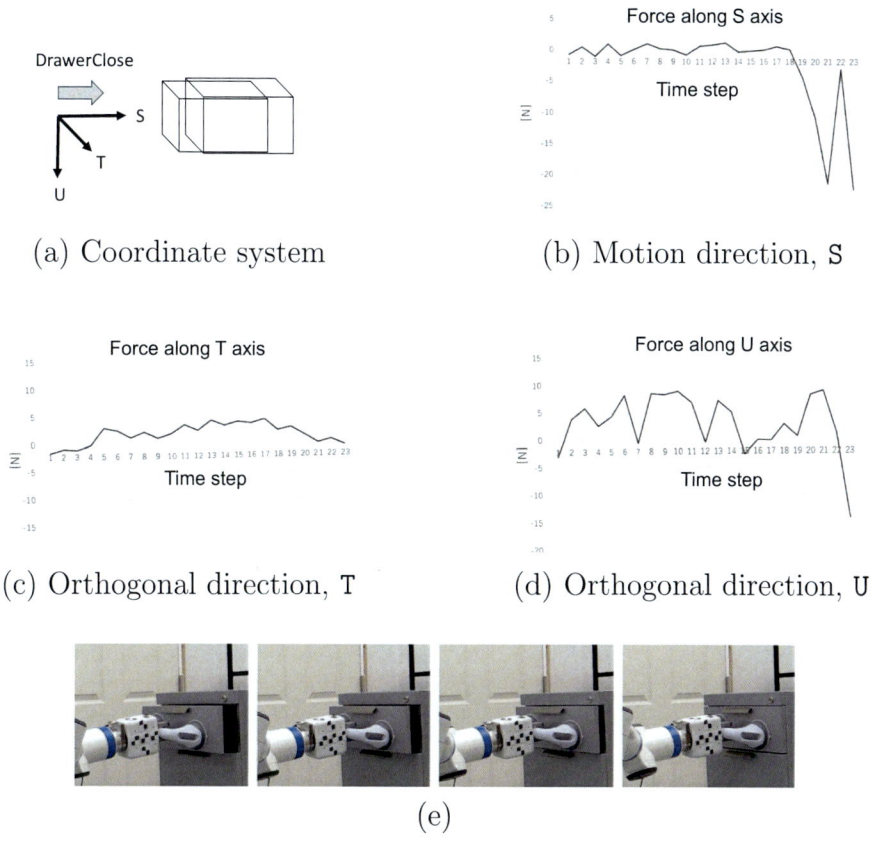

(a) Coordinate system                          (b) Motion direction, S

(c) Orthogonal direction, T                    (d) Orthogonal direction, U

(e)

**Fig. 4.19** Execution of pre-trained `DrawerClose` skill agent. **a** Coordinate system. **b** The magnitude of force along the motion direction, S. **c** The magnitude of force perpendicular to the drawing direction, T. **d** The magnitude of force perpendicular to the drawing direction, U. **e** Robot execution of `DrawerClose` task by the agent

### 4.6.3  Implementing `DoorOpen`/`Adjust`/`Close` skill agents

Similarly to the `Drawer` group, the `Door` group requires reinforcement learning to adjust the rotation direction. These agents are provided with hint information regarding the position of the rotation axis from the demonstration. However, the estimated position of the rotation axis from the demonstration contains errors. Therefore, these skill agents also need to be trained for the adjustment of rotation direction using reinforcement learning.

These agents are trained to perform the rotational motion with an erroneous axis position by learning the directional adjustment of small linear movements. The overall trajectory can be considered as a series of small movements with gradually changing directions. At each step, a small displacement is performed along the tangential movement calculated from the current position with respect to the rotation axis. The pre-training of the agent involves learning a policy that minimizes the resistance force in the direction perpendicular to the target direction. Namely, the pre-training process utilizes the following reward function:

```
if F_b > delta-collision, then penalty
if F_c > delta-collision, then penalty
```

where `B` and `C` are current orthogonal directions to the motion direction `A`.

In the implementation, similar to the `DrawerAdjust` case, the agent is provided with the current motion direction $\mathbf{A}^t$ and the normalized vector representation of the vertical resistance force $\frac{\mathbf{f}^t}{|\mathbf{f}^t|}$ as state values. The action value adjusts the motion vector $\mathbf{A}^t$ with $\Delta\mathbf{A}^t$.

$$\mathbf{A}^{t+1} = \frac{\mathbf{A}^t + \Delta\mathbf{A}^t}{|\mathbf{A}^t + \Delta\mathbf{A}^t|}.$$

The distinction between the `DoorAdjust` and `DrawerAdjust` skills lies in the requirement for the `DoorAdjust` skill to rotate the hand along the tangential direction at each iteration. The update of motion, $\mathbf{A}^{t+1}$, only indicates changes in the hand's position resulting from translational motion. In the case of the `Door`, due to the rotational motion, the hand's grasp direction must be adjusted tangentially at each position. The angle of this hand rotation can be determined from the difference between $\mathbf{A}^{t+1}$ and $\mathbf{A}^t$. This rotation to the grasp direction is added after the hand's translational motion is completed.

For the `DoorOpen` skill agent, the end position can be used as the termination condition, similar to the `DoorAdjust` skill agent. For the `DoorClose` skill agent, the termination condition is when the resistance force in the direction of movement exceeds a certain threshold.

Figure 4.20a shows the execution of the `DoorOpen` skill agent using a real robot, following this pre-trained policy. The displacement distance is set to 5 mm per step. It is worth noting that, in order to alternate motion modification between translational and rotational movements, the grasp must exhibit sufficient flexibility by employing passive form closure rather than passive force closure.

Figure 4.20b illustrates the trajectory of the hand point. The robot attempts to open the door along a curved path, and it can be observed that the estimated opening direction is adjusted to follow the tangential direction of the trajectory. However, due to observational errors in the resistance force direction, the trajectory is not a perfect arc.

### 4.6.4   Implementing `Wipe` skill agent

The `Wipe` task group encompasses three tasks: `WipeStart`, `WipeAdjust`, and `WipeEnd`. For the `WipeStart` and `WipeEnd` tasks, both the constraints on the starting direction and the ending direction are semantic constraints. These constraints are represented using positional information, similar to the `WipeAdjust` task. Thus, from an implementation perspective, they are all handled in the same manner as `WipeAdjust`. Consequently, for the sake of simplicity, they are collectively referred to as the Wipe task.

The reward function for the pre-training of the Wipe skill agent is presented below.

```
Reward function for wipe agent
    if F_t > delta-collision, penalty
    if F_t < delta-zero, penalty
    if F_b > delta-colllsion, penalty
    if F_c > delta-collision, penalty
    if S = goal_s AND U = goal_u AND A = goal_a, reward
```

The reward function specifies that the normal force `F_t` must be less than `delta-collision` and greater than `delta-zero`. This is accomplished by controlling $\mathbf{f}_t$ around an target value $\mathbf{f}_g$.

The state values input into the reinforcement learning algorithm are the surface normal vector, $\mathbf{n}$ and the unit error vector $\mathbf{f}_e$. The unit error vector $\mathbf{f}_e$ can be expressed as follows:

$$\mathbf{f}_e = \frac{\mathbf{f}_t - \mathbf{f}_g}{|\mathbf{f}_t - \mathbf{f}_g|}. \tag{4.12}$$

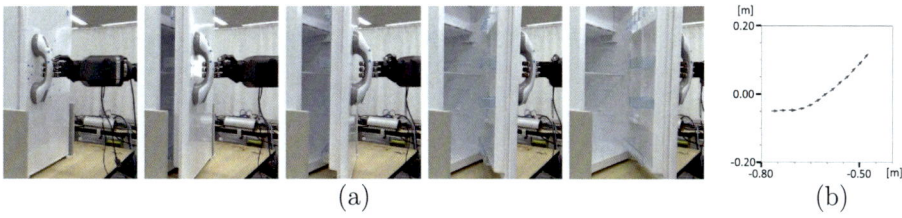

(a)                                                                    (b)

**Fig. 4.20** Execution of pre-trained `DoorOpen` skill agent. **a** Robot execution of `DoorOpen` task by the agent. **b** Moving directions (directions by the directions of the arrows) and the positions at each iteration (marked by the starting points of the arrows) of the robot hand

The reason for using this unit vector instead of the actual force sensor readings is to reduce the Sim2Real gap.

In order to compensate for the effects of gravity, specifically the weight of the hand, the target force $\mathbf{f}_g$ is calculated using the following equation:

$$\mathbf{f}_g = \mathbf{f}_0 + \frac{1}{2} \epsilon \, \mathbf{n}. \tag{4.13}$$

Here, $\mathbf{f}_0$, the force sensor value immediately after contact, can be considered as delta-zero, and $\epsilon = \delta_{collision} - \delta_{zero}$.

The reward function for pre-training is given as:

$$r = \begin{cases} -r_{max} & f > f_{collision} \\ -r_{max} & f < f_{zero} \\ r_{max}/2 - f & otherwise \end{cases}, \tag{4.14}$$

where

$$f = \lfloor \frac{|\mathbf{f}_t - \mathbf{f}_g|}{f_{step}} \rfloor, \tag{4.15}$$

$$f_{collision} = \lfloor \frac{\epsilon}{2 f_{step}} \rfloor, \tag{4.16}$$

$$f_{zero} = \lfloor \frac{-\epsilon}{2 f_{step}} \rfloor. \tag{4.17}$$

While the parameters $f$, $f_{collision}$, and $f_{zero}$ are only utilized during pre-training and do not necessarily require discretization to avoid the Sim2Real gap, they were digitized as a precautionary measure to ensure consistency and mitigate potential issues with the Sim2Real gap. The first and second conditions represent semantic constraints. The third condition guides toward ideal movement, teaching the force to bring the force closer to the target force. The episode's termination condition occurs when either the first or second condition arises, the penalty reaches $r_{max}$, or a predefined period has elapsed. When this predefined period is reached, it is considered successful, and an additional reward of $f_{max}/2$ is given.

The constraint on the direction of the rotation axis,

```
if F_b > delta-colllsion, penalty
if F_c > delta-collision, penalty
```

derived from semantic constraints, was managed by the robot's own rotation axis control.

Figure 4.21 shows the reward curve. The training concluded after 2 million steps.

Figure 4.22a illustrates the execution of the Wipe skill agent using a real robot. The displacement at each step is 5 mm in the direction of movement, with the correction amount calculated from the policy added. Figure 4.22b presents the output from the force sensor

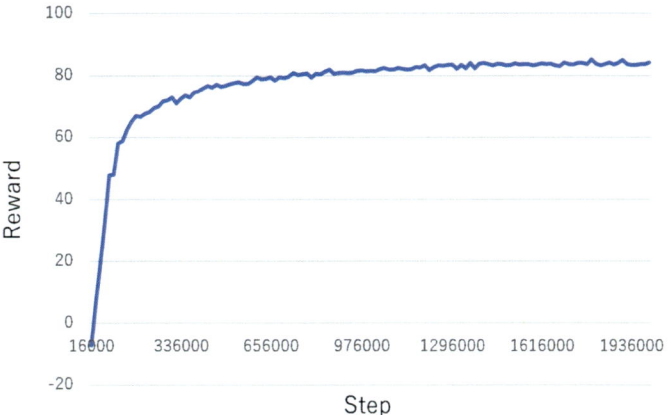

**Fig. 4.21** Learning curve of wipe skill

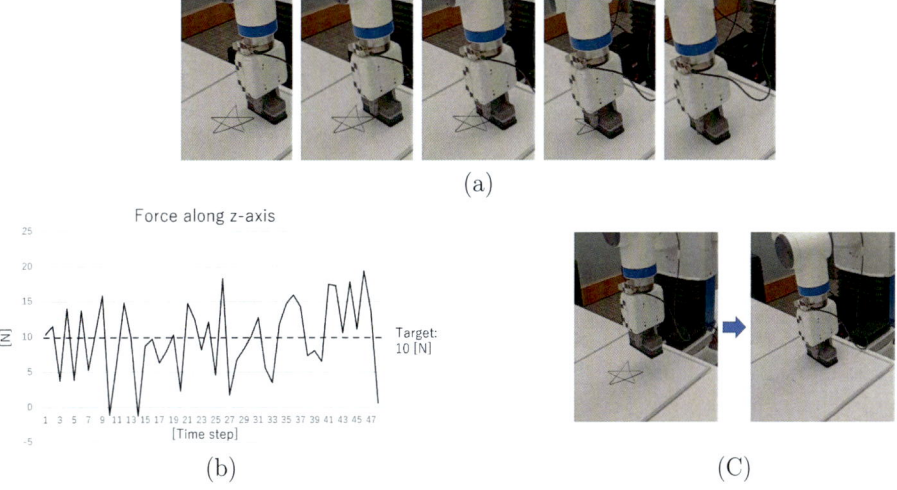

**Fig. 4.22** Execution of `Wipe` task. **a** Robot actions. **b** Force sensor output during the execution. **c** Successfully erase the marking of the white board

during execution, demonstrating that the action is adjusted to achieve the target force of 10 N. As depicted in Fig. 4.22c, the `Wipe` skill agent effectively erased the drawing on the whiteboard.

## 4.7    Related Work

### 4.7.1    Skill Acquisition

The manipulation skills examined in this chapter predominantly encompass tasks performed while accounting for environmental resistance, commonly referred to as compliant manipulation. Historically, enabling robots to execute compliant manipulation has relied on systems designed through classical control theories [136–138]. The practical implementation of these control theories often necessitated extensive manual adjustments of various control parameters, posing a significant and persistent challenge. In recent years, machine learning techniques, such as reinforcement learning and imitation learning, have garnered increasing attention as promising approaches for automating the acquisition of these skills. These approaches aim to train a single neural network capable of independently acquiring the required operational competencies.

Notable advancements in Convolutional Neural Network (CNN) research have driven the emergence of deep reinforcement learning, resulting in significant performance improvements. Many of the policies derived from these advancements have been successfully validated in real-world environments. Examples of successful skill acquisition achieved through skillful reward design in reinforcement learning include studies such as [129, 139–144]. However, the majority of learning processes thus far have been predominantly tailored to single-task specialization. To address this limitation, the field of multi-task reinforcement learning has emerged, focusing on the acquisition of multiple tasks. Despite its promise, this approach faces significant challenges, including the complexity of reward design and the lack of scalability concerning the number of target tasks [145].

Imitation learning, which seeks to replicate human demonstration data, has demonstrated significant success in enabling skill acquisition through supervised learning applied to such data in numerous cases [146–150]. The introduction of the action chunking transformer [86] has further advanced the field by enabling the execution of tasks requiring fine-grained manipulation, characterized by detailed and precise actions. This innovation has attracted attention for its capability to facilitate the acquisition of complex tasks through imitation learning [151, 152].

Imitation learning, which emphasizes trajectory-based learning for individual actions, has encountered significant challenges in its applicability to varying scenarios. Recent advancements in large-scale data learning have fostered progress in developing methods for training a single neural network capable of acquiring diverse tasks [88, 153–155]. Nevertheless, it has been observed that, while the learned models exhibit generalization within the distribution of the training data, their performance significantly diminishes when faced with out-of-distribution actions [155]. Furthermore, even within the scope of the training data, the generalization often enables approximate task execution but falls short in achieving high precision, necessitating fine-tuning to address this limitation effectively.

These studies predominantly emphasize advancing the methodologies of reinforcement learning and imitation learning. However, they lack consideration from the viewpoint of constructing a skill library, as proposed in this chapter. Specifically, there is no discussion regarding the selection of skills to be included in such a library or the appropriate structuring of reward relationships among these skills.

## 4.7.2   Skill Library

To enhance operational accuracy, methods have been proposed that divide the entire task into discrete action units. These units are then used to learn reusable skills, which can be combined to enable general execution across diverse tasks. Such methods rely on multiple predefined reusable skills executed in sequence, guided by plans devised by a task planner. Task planners span a broad range, including classical approaches such as PDDL [156–158], policies learned through hierarchical reinforcement learning [159–162], task planning generation via Diffusion [163], and task planning using large-scale language models [30, 33, 164]. For instance, Brohan et al. [86] defined a skill library by segmenting tasks into discrete action units such as `Pick`, `Place`, `Open drawer` and `Close drawer`, and `Navigate`, with employing the large-scale language model PaLM[165] to carry out task planning. Nasiri-any et al.[166], constructed a skill library by segmenting tasks within environment such as Metaworld [166], Kitchen [167], and Robosuite [168] into four fundamental action units: `Grasp`, `Lift`, `Push`, and `Twist`. They utilized policies learned through hierarchical reinforcement learning to carry out task planning. These approaches demonstrate the advantage of enabling generic execution of various tasks through the combination of learned skills. However, they are limited by the inability to handle scenarios requiring unprepared skills, leading to failures in task execution due to the ad hoc application of task-specific skills.

The methods presented in this chapter differ from the aforementioned approaches in that they aim to address these issues by preparing a set of skills that are both necessary and sufficient within the domain. These methods involve defining each skill along with its generation process in a way that allows the skills to complement one another, ultimately establishing a standardized set of skills.

# Part II
# Restospective Overview of Past LfO

# Big Bang of LfO

5

## 5.1  Object Recognition and Task Recognition

One of the central issues in Learning from Observation (LfO) 2.0, as discussed in the preceding chapters, involves the recognition of human demonstrations through the use of intermediate representations known as task models. Exploring intermediate representations across various task domains has been a common endeavor within LfO. The model-based design of LfO is inspired by the hypothesis that humans perceive the external world through pre-constructed models or templates [5, 169].

In the following chapters, we will examine different intermediate representations within LfO across various task domains. However, before delving into these, we will briefly assess the underlying hypothesis here.

Figure 5.1a illustrates Marr's hypothesis of human object recognition [169]. According to this hypothesis, humans possess abstract object models in their brains, and by associating the features of these models with the features of the external world, they identify objects in the external world. This process is referred to as indexing. They then form a world model in their brains that mimics the spatial arrangement of these objects, a process referred to as localization.

The study of object recognition in computer vision can be considered an exploration of how to obtain this internal model on a computer. Traditionally, before the advent of neural networks, the internal model was obtained by using mathematical formulas or manually extracting features from the object's CAD model or images [99, 170–173]. With the emergence of neural networks, the mainstream approach has shifted to providing large amounts of object images to neural networks and utilizing their learning capabilities to express this

K. Ikeuchi et al., *Learning-from-Observation 2.0*, Synthesis Lectures on Computer Vision, https://doi.org/10.1007/978-3-032-03445-8_5

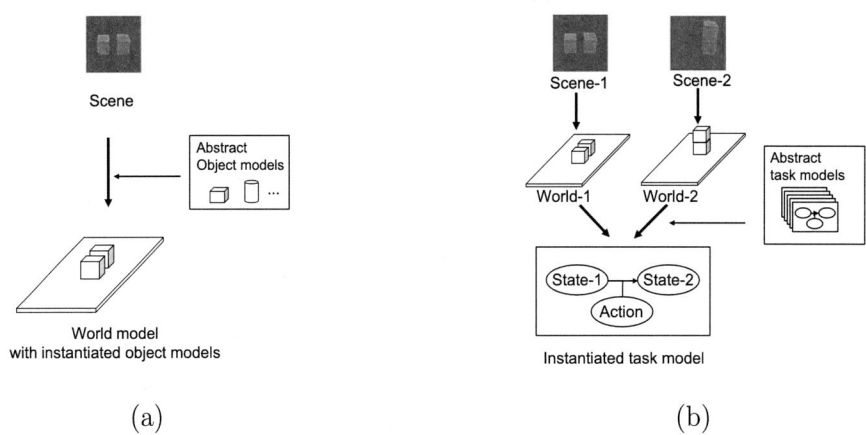

**Fig. 5.1** Object recognition and Task recognition. **a** Object recognition. **b** Task recognition

internal model as the weights of the neural network [174–176]. Regardless of the methodology, the ultimate goal of object recognition research–to obtain this internal model–remains the same.[1]

LfO, which employs task model-based action recognition, extends this concept further (see Fig. 5.1b). Assuming that the previous and current world models are obtained from object recognition, LfO prepares a task model to explain the event that occurred between the two. The structure of this task model is based on Minsky's frame concept [5]. Subsequently, skill agents, which can perform the necessary units of actions to cause the transitions, are devised in response to these task models. A pioneering example of a system that generated robotic actions utilizing this approach is the LfO system developed in 1989 by Reddy and Ikeuchi for the rearrangement of two blocks [7].

## 5.2   Two Blocks World

In 1989, Reddy and Ikeuchi constructed a system that mimicked the human assembly of two blocks [7]. Regarding the assembly of two blocks, only four states exist, as illustrated in Fig. 5.2. Eight transitions among these four states were defined as necessary and sufficient. Consequently, they prepared eight task models, linking the transitions to the requisite actions

---

[1] In this sense, current NN-based object recognition primarily focuses on the initial indexing phase, with fewer studies extending to the subsequent localization phase. This may be related to the fact that, in the human brain, indexing and localization are processed through different pathways. Specifically, indexing occurs through the relatively newer cerebral cortex, aligning well with large language models (LLMs), whereas localization involves older circuits associated with the motor area, potentially requiring unknown computational methods.

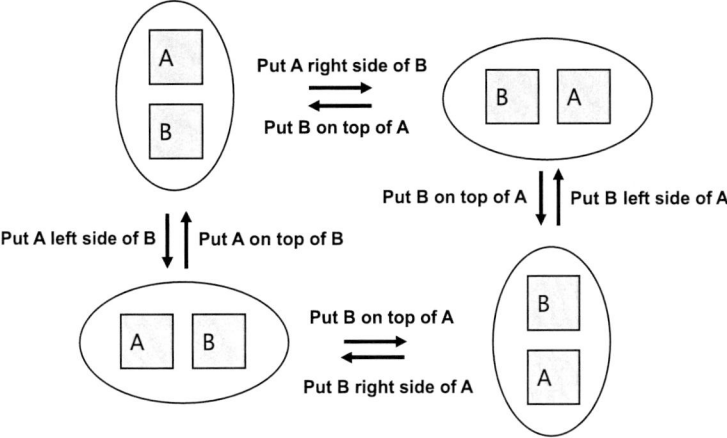

**Fig. 5.2** Four states, eight transitions and eight motion primitives in the two block world

that caused these transitions. After developing the recognition system based on the task models and the robot execution system, the rearrangement of the blocks was performed. They verified that the system successfully mimicked the demonstrator's rearrangement of the blocks and named it "Learning-from-Observation" [8].

The system performed action recognition in a stop-and-go demonstration. Each time the demonstrator performed a unit of actions to replace a block, they moved their hand out of the camera's field of view after completing the actions. The recognition system monitored disturbances in the continuous images. When the hand performed a unit of actions, a disturbance in brightness occurred, and once it was completed, the disturbance decreased. By activating the range sensor when this disturbance decreased and recognizing the blocks based on the depth image, the system could determine the state of the blocks after the actions. By comparing the state of the blocks before and after the actions, the system could identify which state transition (a branch in Fig. 5.2) occurred. From this, the system retrieved the necessary action associated with that branch to perform the rearrangement of the blocks as shown in the YouTube video.[2]

This two-block system comprises three modules. The first module activates the range sensor based on disturbances in brightness to obtain depth images before and after a unit of actions, followed by object recognition based on the pair of depth images. The second module retrieves the task model derived from the object recognition results, aligning with the current task encoder that utilizes a Large Language Model (LLM) in LfO2.0. This module within the two-block system is aware of the transition diagram shown in Fig. 5.2, analogous

[2] https://youtu.be/UkvDUmRHg4c.

to the LfO2.0 system, which is provided with the set of skill agents available in the skill libraries as the prompt. Additionally, while the robot execution system in this two-block system was implemented in an ad hoc manner, the LfO2.0 system employs agents utilizing reinforcement learning.

Although the implementation of this two-block system was rudimentary, it laid the foundation for subsequent systems. The concept of pairing state transitions with the units of actions that cause them, known as task models, became the central focus of later research in LfO. Since then, LfO research has involved defining various domains of human tasks and establishing sets of task models within those domains that satisfy necessity and sufficiency.

# Poloyhedral World

**6**

This chapter examines the methodologies for defining task models based on contact transitions in the polyhedral world, as well as the resulting LfO implementation based on these definitions, as presented by Ikeuchi and Suehiro [9]. Similar to the two-block world discussed in the previous chapter, the system is composed of four key processes: image sequence segmentation via a stop-and-go approach, object recognition, task recognition, and task execution. This chapter primarily focuses on the definition and recognition methodologies of the task models.

Ikeuchi et al. enumerated all the task models in the world of polyhedra. Although the set of task models in the two-block world was similarly defined, it did not cover all possible contact states of the two blocks and, therefore, did not account for all task models. To address this limitation, they clarified the definition of contact states and enumerated all possible transitions for setting up the task models. This concept of enumerating all possible transitions became a fundamental design principle for subsequent Learning from Observation (LfO) systems in various domains.

We define various terminologies to describe the assembly process of polyhedrons. In each assembly operation, at least one object is manipulated. This object is referred to as the *manipulated object*. The manipulated object is attached to other stationary objects, referred to as the *environmental objects*, and the manipulated object achieves a contact relation with the environmental objects, defined as the assembly relationship.

The main goal of the assembly operation is to establish a new assembly relationship. For example, the goal of peg insertion is to achieve surface contact between the sides and bottom of the peg (manipulated object) and the sides and bottom of the hole (environmental object). We use the assembly relationship and its transitions to define the assembly operations.

An assembly relationship describes only the contact relationships between the manipulated and environmental objects. It does not account for other contact relationships among different objects. By focusing solely on the relationship between the manipulated object and the environmental objects, we prevent combinatorial explosion. Furthermore, to simplify the descriptions, we consider only the manipulation of one polyhedral object in each assembly operation.

Since the assembly relationships meet the requirements of recoverability and inferability, as outlilned below, we will use these relationships as the fundamental representations to describe assembly operations.

**Recoverability: Can assembly relationships be extracted from observation?** The object recognition program identifies the manipulated object, determines its placement, and utilizes a geometric modeler to represent the recognition results. Additionally, the geometric modeler represents all other stationary environmental objects. By examining each pair of faces of the manipulated object and the environmental objects using the representations provided by the geometric modeler, it is possible to determine the surface contact relationships, and thus, their assembly relationships between these objects.

**Inferability 1: Can human assembly operations be inferred?** Assembly relationships consist of multiple surface contact relationships, which constrain the possible movements of the manipulated object. On the contacting surfaces, the orientation of the surface normals is sufficient to characterize the constraints on the relative movement of the object. For example, if a box (the manipulated object) is placed on a table (the environmental object), the normals of the contact surfaces are parallel and facing each other, allowing the box to move only upward or parallel to the table. A more constrained case occurs when a square rod (the manipulated object) is inserted into a hole of the same shape (the environmental object). The four surfaces of the rod are in contact with the corresponding surfaces of the hole, with opposing normals, permitting movement only along the axis of the hole. Therefore, by analyzing assembly relationships, it becomes feasible to deduce the assembly operations encompassed within the potential movements of the manipulated object, leading to the observed assembly relationships.

**Inferability 2: Is it possible to reproduce this with a robot?** Each surface contact relationship, as a component of the assembly relationship, constrains the possible movements of the manipulated object. As long as the constraints on movement remain constant, the same control mode can be applied. However, when the constraints on movement change, different control modes are required. Thus, each surface contact relationship determines the necessary control mode. For instance, consider a scenario where a box is first placed on a table and then slid across it. When the box is in the air (with no surface contact), position control can be utilized. As the box approaches contact with the table (when a single surface contact is about to occur), force control is needed to detect the collision and confirm that the box is on the table. To slide the box across the table, a combination of position control and force control is required to maintain the single surface contact. In this manner, surface contact relationships can be used to determine the control strategies necessary to achieve and maintain such surface contact relationships.[1]

We will classify all possible assembly relationships, consider the types of transitions that occur within these assembly relationships, and construct a graph where each node corresponds to an assembly relationship and each branch represents a possible transition. The task model will be designed by assigning the units of actions necessary to achieve such transitions to each branch.

## 6.1  Classification of Assembly Relation

This subsection enumerates possible classes of assembly relationships in the context of translational operations [18]. According to the Kuhn-Tucker theory, assembly relationships can be classified into ten distinct classes based on the characteristics of solution spaces for simultaneous linear inequalities. To explore the relationship between Kuhn-Tucker theory and our assembly relationships, we examine the constraints on the motion of polyhedrons by considering the rank of simultaneous inequalities that represent these constraints. Our investigation yields classification results consistent with those predicted by the Kuhn-Tucker theory. Hence, this subsection demonstrates Kuhn-Tucker theory using face contacts as concrete examples.

### 6.1.1  A Constraint Given by a Surface Pair

Suppose the surface patch of the manipulated object is in contact with the surface patch of the environmental objects. This surface contact pair constrains the possible translations of the manipulated object as follows:

$$\mathbf{N} \cdot \mathbf{X} \geq 0, \tag{6.1}$$

where $\mathbf{X}$ denotes a possible translational motion vector of the manipulated object and $\mathbf{N}$ denotes the normal direction of an environmental surface patch.

Using points on the Gaussian sphere, we can specify constraint unit vectors as well as all possible translation vectors. Each unit vector is positioned such that its starting point is at the center of the Gaussian sphere, with its endpoint on the sphere's surface. This endpoint uniquely identifies the direction of the vector, allowing us to represent the vector by this point. By utilizing the entire spherical surface, we can denote all possible directions of the constraint unit vector.

---

[1] This original discussion leads to the dimensional analysis for designing the reward functions described in Sect. 4.3.

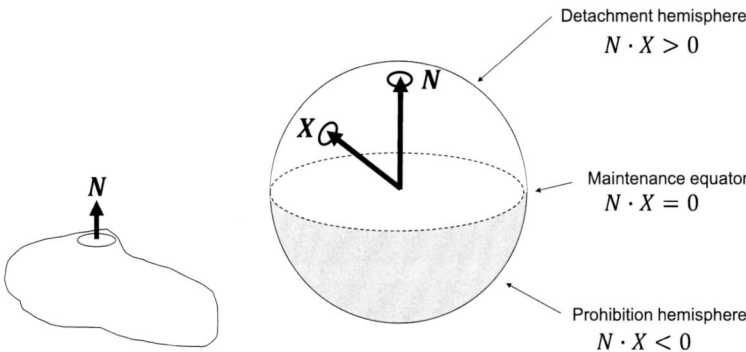

**Fig. 6.1** One-directional contact and Gaussian sphere

We can assume, without loss of generality, that the constraint unit vector **N** is oriented towards the north pole of the Gaussian sphere. Consequently, the normal of the previous surface can be represented as the north pole of the Gaussian sphere.

Constraints due to surface contacts define several regions on the Gaussian sphere. The plane perpendicular to the normal vector **N** is called the constraint plane, and this plane divides the Gaussian sphere into two hemispheres. Points in the northern hemisphere (called the detachment hemisphere) satisfy $\mathbf{N} \cdot \mathbf{X} > 0$, indicating motion vectors that break the surface contact. Points in the southern hemisphere (called the prohibited hemisphere) satisfy $\mathbf{N} \cdot \mathbf{X} < 0$, indicating prohibited movements where the manipulated object collides with the environment objects. Points on the equator (called the maintenance equator) satisfy $\mathbf{N} \cdot \mathbf{X} = 0$, corresponding to movements that maintain surface contact. See Fig. 6.1.

### 6.1.2  Unidirectional Constraint Relation

Consider a scenario where all contact patch pairs share the same direction but differ in their physical positions. The constraint inequalities given by all patch pairs have the same coefficient $\mathbf{N_1}$ and yield exactly one constraint inequality equation.

$$\mathbf{N_1} \cdot \mathbf{X} \geq 0. \tag{6.2}$$

This assembly relationship is labeled as the PC1 (Partial Contact 1) relationship. Each assembly relationship provides certain motion directions that do not cause collisions between the manipulated object and the environmental objects (allowable motion directions) as determined by the constraint equations. The allowable motion directions for the PC1 assembly relationship form a single angular region (hemisphere) on the Gaussian sphere, including its boundary, which is the union of the detachment hemisphere and the maintenance equator.

Motion directions in the boundary region (along the equator) maintain the PC1 assembly relationship. These motion directions are called the maintenance motion directions of the PC1 relationship. The degree of freedom for the maintenance motion directions, the maintenance DoF, is 2.

Motion directions corresponding to the interior region (within the northern hemisphere) disrupt the PC1 relationship and are termed detachment motion directions of the PC1 relationship. Any detachment motion can be represented as a spherical convex set of maintenance and pure detachment motions. Here, a pure detachment motion refers to one that does not include maintenance motion components. Pure detachment motions of the PC1 relationship are aligned with the constraint normal vector $\mathbf{N_1}$, and the degree of freedom for pure detachment, the detachment DoF, is 1.

### 6.1.3   Bidirectional Contact

Bidirectional contact provides the following two constraint inequality equations.

$$\mathbf{N_1} \cdot \mathbf{X} \geq 0, \tag{6.3}$$

$$\mathbf{N_2} \cdot \mathbf{X} \geq 0. \tag{6.4}$$

The rank of the coefficient matrix, given by the inequality equations for $\mathbf{N_1}$ and $\mathbf{N_2}$, is either 1 or 2.

#### 6.1.3.1 Rank $= 2$

The two constraint planes defined by $\mathbf{N_1}$ and $\mathbf{N_2}$ intersect to form an allowable crescent-shaped biangular region on the Gaussian sphere. This relationship is referred to as the PC2 (Partial Contact 2) relationship. The allowable biangular region for the manipulated object is bounded by two semicircles of great circles and two vertices. Refer to Fig. 6.2a for details.

The points at the two vertices satisfy both $\mathbf{N_1} \cdot \mathbf{X} = 0$ and $\mathbf{N_2} \cdot \mathbf{X} = 0$. The motion directions corresponding to these points maintain this PC2 relationship. The maintenance DoF is 1.

Points along one of the semicircles correspond to motion directions that maintain one of the contact relationships and break the other.

Motion directions corresponding to points inside the boundary are detachment motion directions. Pure detachment motion directions correspond to points along the arc connecting $\mathbf{N_1}$ and $\mathbf{N_2}$ on the Gaussian sphere, and the detachment DoF is 2.

### 6.1.3.2 Rank = 1

When the rank is one, the two normal vectors must either have the same or opposite directions. If the two vectors have the same direction, it conflicts with the definition of bidirectional contact. Therefore, the two vectors should have opposite directions. Namely,

$$\mathbf{N_1} \cdot \mathbf{X} = 0,$$
$$\mathbf{N_2} \cdot \mathbf{X} = 0,$$
$$-\mathbf{N_1} = \mathbf{N_2}.$$

These two directions can be represented as a pair of poles on the Gaussian sphere, such as the north pole and the south pole. The allowable directions for the manipulated object can be represented as the entire great circle perpendicular to the axis connecting the two poles.

This assembly relationship is referred to as the TR (Translational Contact) relationship. Refer to Fig. 6.2b. Each allowable motion direction in the TR relationship satisfies the two aforementioned equations. All allowable directions of the TR relationship maintain the TR relationship, and the maintenance DoF is 2. There are no detachment motions, and the detachment DoF is 0. One direction along the axis connecting $\mathbf{N_1}$ and $\mathbf{N_2}$ is completely constrained and is referred to as the constraint direction. The constraint DoF is 1.

## 6.1.4  Tridirectional Contact

Tridirection contact provides the following three constraint inequality equations.

$$\mathbf{N_1} \cdot \mathbf{X} \geq 0,$$
$$\mathbf{N_2} \cdot \mathbf{X} \geq 0,$$
$$\mathbf{N_3} \cdot \mathbf{X} \geq 0.$$

The rank of this coefficient matrix is either 3, 2, or 1.

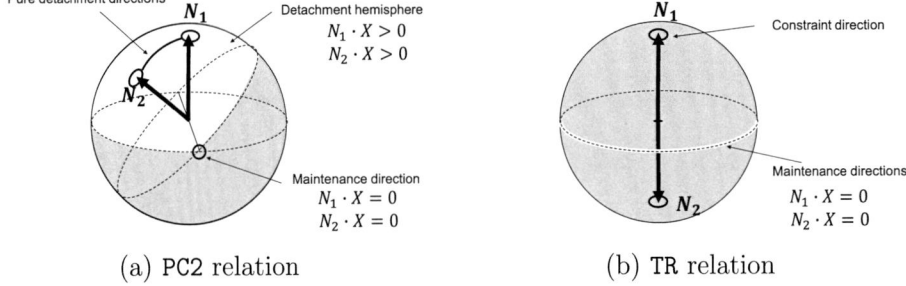

(a) PC2 relation                         (b) TR relation

**Fig. 6.2** Two directional contacts

### 6.1.4.1 Rank 3

The three constraint planes intersect at the center of the Gaussian sphere, forming an allowable triangular region on the Gaussian sphere. See Fig. 6.3(a). This relationship is referred to as the PCN (Parietal Contact N-polygon) relationship.

All allowable motion directions corresponding to the interior region are detachment motions, which break the contact relationships. Motion directions along the boundary of the triangle break two of the three contact relationships while maintaining the remaining one. Motion directions at the vertices of the triangle break one contact relationship while maintaining the other two. Therefore, the detachment DoF in the PCN relationship is 3, while both maintenance DoF and constraint DoF are 0.

### 6.1.4.2 Rank 2

The three normal vectors are coplanar. In this case, it can be further divided into two subcases.

- *All pairs are independent*: If all pairs of vectors are independent (i.e., the rank of all submatrices provided by the two constraint inequality equations is 2), the three vectors form either an allowable biangular region (equivalent to a PC2 relationship, see Fig. 6.3b) or a pair of allowable poles with an axis perpendicular to the coplanar vectors (PR relationship (PRismatic contact), see Fig. 6.3c).
  In the PR relationship, all allowable motions are maintenance motions along the axis connecting a pair of the allowable poles, and the maintenance degree of freedom is 1. The constraint direction is in the coplanar plane, and the constraint DoF is 2. Since there are no detachment motions in the PR relationship, the detachment DoF is 0.
- *One pair forms polar opposites*: If one of the two vectors is in the opposite direction (i.e., one submatrix has rank 1), this pair of the two vectors forms an allowable great circle perpendicular to them. See Fig. 6.3d.

The third vector is linearly independent of the first two vectors and divides the entire great circle into a semicircle. Therefore, the semicircle represents the allowable directions. This is called the OT1(One-way Translation 1) relationship.

The maintenance motion directions correspond to both ends of the semicircle, and the maintenance DoF is 1. Pure detachment motions are along the direction of the third vector, and the detachment DoF is 1. The pair of opposite direction vectors provides the constraint direction, and the constraint DoF is 1.

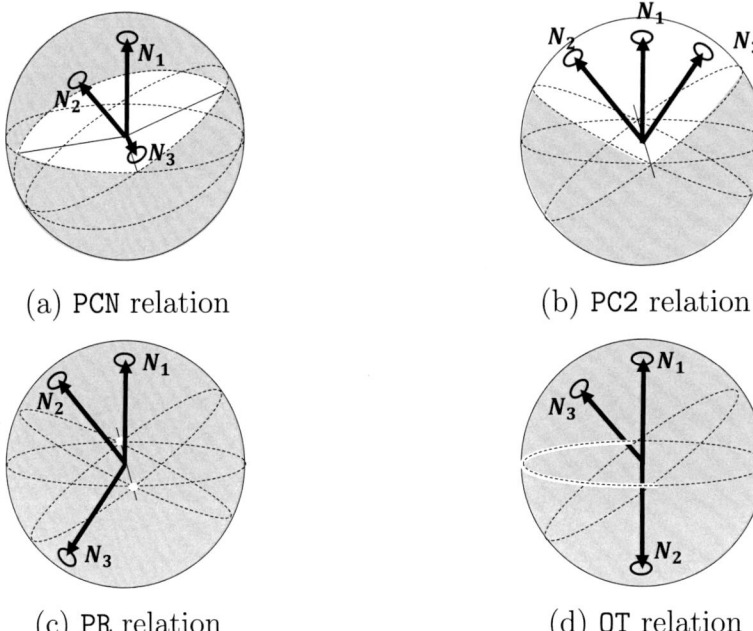

(a) PCN relation                                    (b) PC2 relation

(c) PR relation                                     (d) OT relation

**Fig. 6.3** Tridirectional contact

### 6.1.4.3 Rank 1
Due to the definition of three-direction contact, this case does not occur.

## 6.1.5   Tetradirectional Contact

Tetradirectional contact provides four constraint inequalities. To analyze this case, we use an inductive approach that considers how the additional constraint vector affects the allowable region formed by the first three vectors.

### 6.1.5.1 Triangular Region
When adding the fourth constraint vector $N_4 \cdot X \geq 0$, the intersection with the triangular region formed by the first three vectors can lead to different scenarios, such as:

- *Non-intersection*: Either of the following two cases occurs depending on whether the triangular region lies in the detachment hemisphere or the prohibition hemisphere of the new constraint:

- *Detachment*: If in the detachment hemisphere ($\mathbf{N_4} \cdot \mathbf{X} > 0$), the new allowable region retains the shape of the previous triangular region. In this case, it is the same as the PCN relationship.
- *Prohibition*: If in the prohibition hemisphere ($\mathbf{N_4} \cdot \mathbf{X} < 0$), no allowable region exists. This relationship is labeled as FT (Fully Contact) relationship. In this case, all degrees of freedom are constrained; the constraint DoF is 3.

- *Intersection*: Depending on how the fourth constraint plane intersects the triangular region, the following three cases occur:

  - *General Case*: When the fourth constraint plane intersects the triangular region, the region is divided into two sub-regions. These sub-regions can take on triangular or quadrilateral shapes. See Fig. 6.4a. In particular, a quadrilateral allowable region over-constrains the object from four directions. This has the same effect as the PCN three-direction contact. That is, any three of the four-direction contacts automatically ensure the fourth contact. Therefore, this relationship is treated as equivalent to PCN.
  - *Edge Alignment*: When the fourth constraint plane aligns with one of the edges of the triangular region, the triangular region will be in either the detachment hemisphere or the prohibition hemisphere:

    - *Detachment* If in the detachment hemisphere, the newly added normal direction aligns with the normal direction which forms the edge, and the case becomes three-directional contact and not four-directional contact. Thus, this case contradicts the definition of four-directional contact.
    - *Prohibition* If in the prohibition hemisphere, four-directional contact is established. In this case, the new allowable region is reduced to the arc along the great circle corresponding to the edge of the triangular region. This relationship is labeled as OT2 (One-way translational contact 2) relationship. See Fig. 6.4b. In this case, all the motion directions towards the points on the arc lead to detachment from some contact surfaces, and the detachment DoF is 2. On the other hand, the motion direction is prohibited along the two opposing constraint directions, with the constraint DoF being 1.

  - *Vertex Passage*: Consider the case where the fourth constraint plane passes through one of the vertices of the triangular region. The triangle exists in either the detachment hemisphere or the prohibited hemisphere:

    - *Detachment*: If the triangular region is in the detachment hemisphere, the triangular region completely resides within the detachment hemisphere of the new fourth constraint plane. The new constraint does not affect the allowable spherical region and is equivalent to the PCN relationship.

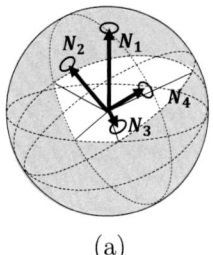 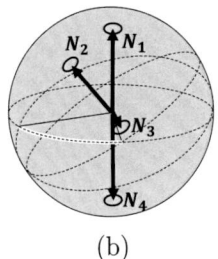 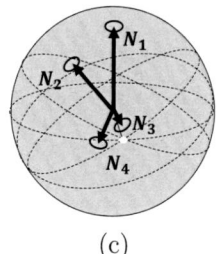

(a)                              (b)                              (c)

**Fig. 6.4** Tetradirectional contact

- *Prohibition*: If in the prohibition hemisphere, the new allowable region becomes only the vertex (point). This relationship is labeled as the OP (One-way Prismatic contact) relationship. See Fig. 6.4c.
  Motion directions toward the point break the contact relationships. Thus, the detachment DoF is 1. The plane perpendicular to the detachment motion direction is the constraint plane, and the constraint DoF is 2. There are no maintenance motions, and the maintenance DoF is 0.

### 6.1.5.2 Biangular Region
This case occurs when the three vectors are coplanar, and two of them form a biangular region (e.g., Fig. 6.3b). Depending on the direction of the fourth vector, one general and singular cases arise.

- *General case*: If the fourth constraint normal is not coplanar with the other three normals, the fourth constraint plane intersects the biangular region and forms a new triangular allowable region. This contact relationship is equivalent to the PCN relationship.
- *Coplanar*: If the fourth constraint normal is coplanar with the other three normals, depending on the direction of the fourth normal, three cases occur.

  - *Detachment*: If the biangular region is in the detachment hemisphere of the new constraint, it remains the same biangular region and is equivalent to the PC2 relationship.
  - *Prohibition*: If in the prohibition hemisphere of the new constraint, only the two vertices of the biangular region maintain the degree of freedom, which is equivalent to the PR relationship.
  - *Maintenance*: If the edge of the biangular region aligns with the maintenance boundary of the new constraint and the biangular region is in the prohibition hemisphere, only the boundary line maintains the degree of freedom, which is equivalent to the OT1 relationship.

### 6.1.5.3 Half of a Great Circle

This is a special case of the aforementioned biangular region. Not only are the three vectors coplanar, but two of them must also be in opposite directions. Essentially, these three constraints form a common detachment motion region equivalent to the OT1 relationship and form a half circle on a great circle. In relation to this circle, there are two possible scenarios depending on how the fourth constraint intersects with this:

- *Non-coplanar*: This is a general case: when the fourth vector is not coplanar with the previous three vectors, the allowable region becomes an arc along a great circle (equivalent to the OT2 relationship).
- *Coplanar*: This is a singular case: when the fourth vector is coplanar, there exists another opposite direction vector that is not part of the previous pair. The fourth vector forms a pair of opposite directions with that vector, creating two pairs of opposite vectors on the coplanar plane. Consequently, the allowable region becomes a pair of poles (equivalent to the PR relationship).

### 6.1.5.4 Pair of Pole Points

As with the previous arc, two scenarios emerge depending on the relationship between the pair of poles generated by the first three constraints and the fourth constraint.

- *General case*: In the general case, depending on the orientation of the fourth constraint's normal, one of the two poles falls into the prohibited hemisphere while the other pole falls into the detachment hemisphere. This results in a single allowable point, equivalent to the OP relationship.
- *Coplanar*: In the singular case where the fourth normal is coplanar with the other normals, the pair of poles remains unchanged and is equivalent to the PR relationship.

### 6.1.6   N Directional Contact

By using a similar inductive method as the four-direction contact relationship, we can prove that these nine relationships are sufficient to describe general directional contacts ($n > 4$).

For completeness, we also include the relationship where there is no surface contact (NC). All motions maintain the same NC relationship, with the maintenance DoF being 3. The constraint DoF and the detachment DoF are zero. Figure 6.5 shows the ten types of relationships.

Similar conclusions can be obtained using the Kuhn-Tucker theory.

**Fig. 6.5** Ten states

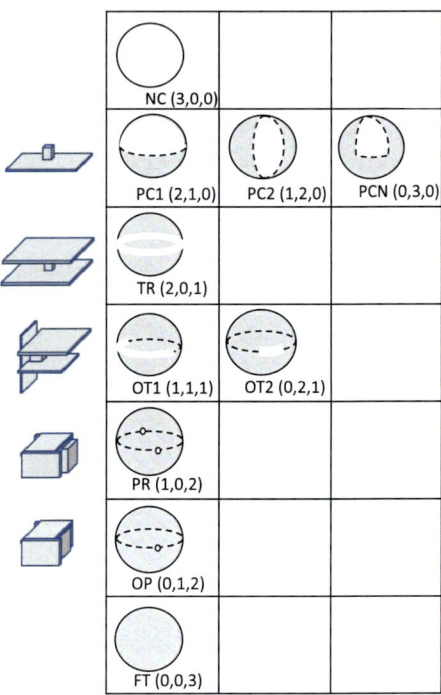

The three values depicted in Fig. 6.5 represent the degrees of freedom associated with maintenance DoF ($D_m$), detachment DoF ($D_d$), and constraint DoF ($D_c$). The sum of these three values consistently totals three, reflecting the overall degrees of freedom for an object in translational motion within 3D space. Notably, the numbers, whose sum is three, are allocated to these degrees of freedom in dictionary order.

$$D_m + D_d + D_c = 3. \tag{6.5}$$

## 6.2   Manipulation Skills

This section determines the sequence of manipulator operations required to achieve each assembly relationship from the NC assembly relationship. By grouping these sequences into operation templates and providing appropriate control parameters, the desired assembly relationship can be generated. This is possible because each assembly relationship is defined in such a way that a common control strategy—requiring only modifications to the control parameters—can be applied to achieve the relationship. A robot operation that triggers a transition into an assembly relationship is called a *manipulation skill*, and a sequence of manipulator control actions that is commonly used among multiple skills is referred to as a *motion macro*.

To reduce the number of required motion macros, intermediate relations are employed. Initially, manipulation skills for achieving simple relations are considered. For more complex relations, rather than directly achieving them from the NC, we aim to achieve known intermediate relations using established skills, and then proceed to achieve the target relation with a new skill from that point. This approach begins with no contact (NC relation) and gradually increases the number of contacts, progressing towards more complex assembly relations.

The difference between the LfO2.0 system described in Part I and the system in this chapter is that while the LfO2.0 considered home service robots, including both increasing and decreasing surface contact directions as manipulation skills, the system in this chapter defines machine assembly only in the direction of increasing surface contact and only considers skills in the direction of increasing surface contact. Additionally, in LfO2.0, when considering all possible transitions, PC1, PC2, and PCN are grouped into one group, and OT1 and OT2 into another, thereby reducing the number of states. In contrast, this system treats these independently. However, the fundamental design philosophy is common to both.

## 6.2.1 Disassembly Analysis

To find the appropriate intermediate relations, we consider disassembly actions (where the number of surface contacts gradually decreases) instead of assembly actions (where the number of surface contacts gradually increases). By considering disassembly actions from an assembly relation, we extract intermediate assembly relations that can be directly reached from that assembly relation. This approach is chosen because disassembly actions are easier to infer than assembly actions.

### 6.2.1.1 Action Selection Rule

Due to differences in the shape of the contact surfaces and the direction of the disassembly action, it is possible to reach multiple intermediate assembly relations from the same assembly relation. In such cases, the intermediate relations are selected or not selected according to the following rules:

- **Shape rule**: In the case of differences in the contact surface shape, all intermediate relations are analyzed, and an appropriate control strategy is assigned to the transition from each intermediate relation back to the original relation.
- **Direction rule**: When multiple intermediate relations are reached due to the direction of the action, the simplest and most robust action that can be achieved even with uncertain positional information is selected using the following criteria:

- **Criterion 1**: if a direct detach action, which immediately breaks the surface contact, select it.
- **Criterion 2**: if a lateral action, which maintains the same contact relation, breaks the surface contact by crossing a specific boundary, select it.
- **Criterion 3**: if multiple candidate actions satisfy criterion 1 or 2, select the action that reduces the number of surface contacts the least.

### 6.2.1.2 Transition Graph

The disassembly analysis from each assembly relation will provide the following transitions:

- **From PC1 relation**: A direct separation action causes the transition from PC1 to NC. See Fig. 6.6a.
- **From TR relation**: No transition occurs due to infinitesimal movements. With finite movements, depending on the shape of the contact surface, it reaches either NC or PC1. Since the change is due to the shape of the contact surface, according to the shape rule, both transitions are included. See Fig. 6.6b.
- **From PC2 relation**: By applying direct separation motion, the PC2 relation reaches NC or PC1 depending on the direction of the motion. The transition from PC2 to NC reduces the number of constraints by two, while the transition from PC2 to PC1 reduces it by one. The latter relationship transition is selected according to criterion 3 in the direction rule. See Fig. 6.6c.
- **From OT1 relation**: A direct detach motion causes the transition from OT1 to TR.
- **From PR relation**: In the case of PR relationship, no transition occurs due to infinitesimal movements. By performing finite maintenance movements that satisfy the constraints from the surrounding contact surfaces, various relations such as NC, PC1, TR, PC2, and OT are induced depending on the shape of the contact surfaces. These five possible transitions are included. See Fig. 6.7.
- **From PCN relation**: PCN reaches NC, PC1, or PC2 depending on the direction of the motion. The transitions to NC, PC1, and PC2 reduce the number of constraints by three,

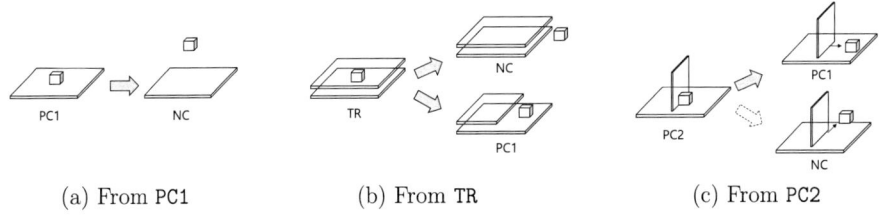

(a) From PC1          (b) From TR          (c) From PC2

**Fig. 6.6** Disassembly analysis

**Fig. 6.7** Disassembly analysis of PR relation

two, and one, respectively. Therefore, according to criterion 3 in the direction rule, the transition to PC2 is selected as the desirable one.

- **From OT2 relation**: Among the two possible transitions to TR and OT1, the transition to OT1 is selected based on criterion 3 in the direction rule.
- **From OP relation**: The only possible motion is a detachment motion along the axis perpendicular to the constraining surface, which leads to TR.

By integrating the possible transitions derived from the aforementioned disassembly analysis, a graph is constructed that illustrates the path from each assembly relation to the NC, utilizing intermediate assembly relations, as depicted in Fig. 6.8. The nodes in the figure denote each assembly relation, while the arcs represent the paths obtained from the preceding disassembly analysis. Conversely, by tracing this graph from the NC, one can navigate to any assembly relation through intermediate assembly relations.

## 6.2.2   Skill Assignment

We will investigate the requisite skills needed to accomplish the assembly relation transitions. Based on the arcs in Fig. 6.8, which indicate the directions of disassembly, we will consider the skills to achieve such transitions along the reverse arcs in the graph.

### 6.2.2.1 Skills for Transitions
- **NC-PC1 skill**: To achieve the transition from NC to PC1, an approach action toward the contact surface of the environmental object is executed. Among the various possible actions, the pure attaching action—where the contact surface is approached from the opposite direction of its normal—is chosen to simplify the process. This action continues until contact with the surface is achieved. This series of movements is termed the move-to-contact motion macro. The NC-PC1 skill, consisting of

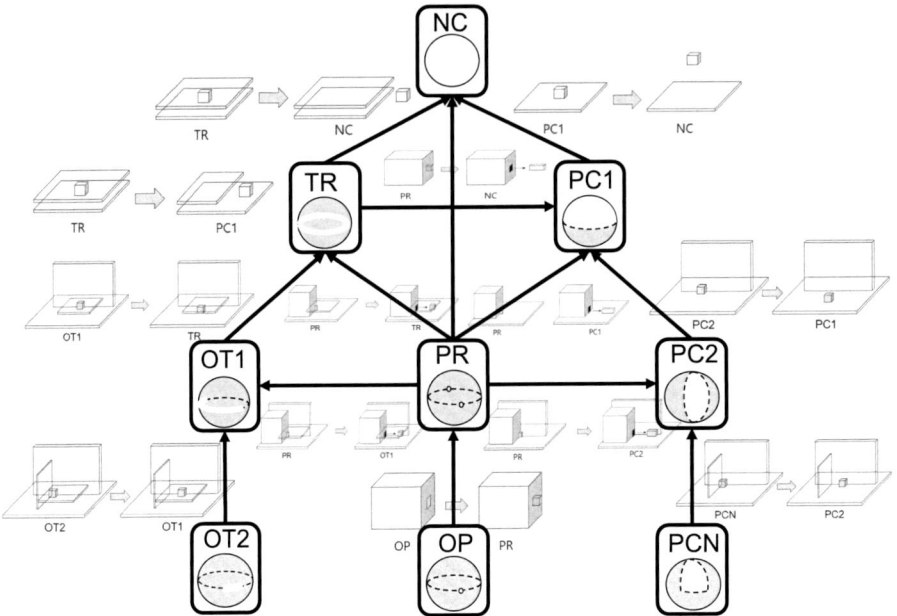

**Fig. 6.8** Disassembly transitions

move-to-contact motion macro, is also sometime referred to as Place skill as a friendly name, and executing the NC-PC1 skill is equivalent to performing this move-to-contact motion macro.

- **NC-TR skill**: To achieve the transition from NC to TR, the object's position and orientation relative to the gap between the two contact surfaces must first be adjusted. Next, the object must be moved along the contact surface. This collection of movements is referred to as the insert-between motion macro. In other words, NC-TR skill can be realized by using insert-between motion macro.
- **PC1-TR skill**: By translating the object along its contact plane until it reaches the target configuration, bidirectional contact occurs automatically. This simple translational action, achieved by move motion macro, constitutes PC1-TR skill.
- **PC1-PC2 skill**: Among the possible actions that maintain the PC1 state, the transition from PC1 to PC2 can be achieved by a collision action towards a new constraint plane. To simplify the process, among various collision actions, the action in the direction perpendicular to the intersection line of the two constraint planes is selected. This action can be achieved using the move-to-contact motion macro.
- **TR-OT1 skill**: The possible actions in the TR state are those along two opposing constraint planes. During this action, by directing movement towards a third constraint plane, a collision occurs, achieving the transition from the TR state to the OT1 state. Similar to

the PC1 to PC2 case, a pure attaching action is selected. Consequently, the same skill as in the `PC1-PC2` transition, namely the `move-to-contact` motion macro, is assigned.

- **NC-PR skill**: To achieve the transition from NC to PR, it is first necessary to align the position and orientation of the object with the insertion axis to enable movement along this axis. Following alignment, the object is moved along the axis. This set of actions is referred to as the `insert-into` motion macro.

  The `insert-between` motion macro used for the NC-TR transition only adjusts the object parallel to a pair of contact surfaces, allowing freedom for rotation and translation along the contact surfaces. In contrast, the `insert-into` motion macro restricts such freedom, permitting the object to move exclusively along the insertion axis.

- **PC1-PR skill**: The maintenance motion in the PC1 state already constrains one of the two degrees of freedom (DOF) required for constraints in the PR state; not deviating from this direction is equivalent to constraining motion along this direction. Consequently, if the transition from PC1 to PR occurs while maintaining the surface contact, the increase in constrained DOFs will be one.

  To accommodate this increase in constrained DOFs, the `insert-between` motion macro is employed. This macro aligns the configuration perpendicular to the detachment normal of the PC1 relation, enabling the object to fit into the hole seamlessly.

- **TR-PR skill**: This can be achieved using the `insert-between` motion macro, similar to the transition from PC1 to PR.

- **PC2-PR skill**: The maintenance action of the PC2 state translates the object along the intersection line of the two contact planes, aligning with the insertion axis. The `move` motion macro is used to perform this transition.

- **OT1-PR skill**: This can be achieved using the `move` motion macro, similar to the transition from PC2 to PR.

- **PC2-PCN skill**: The PCN state can be achieved by performing the maintenance action of the PC2 state until three-directional contact is generated, using the `move-to-contact` motion macro.

- **OT1-OT2 skill**: The maintenance actions of the OT1 state translate the object along the direction connecting the two endpoints of the allowable half of the great circle on the Gaussian sphere. The `move-to-contact` motion macro is used for this maintenance action until four-directional contact occurs.

- **PR-OP skill**: The transition from PR to OP is achieved using the `move-to-contact` motion macro along the insertion axis, which is the maintenance action of the PR relationship, until four-directional contact occurs.

### 6.2.2.2  Specifications of Motion Macros

Based on the aforementioned discussion, the following four motion macros have been identified as essential components for implementation of the skills:

**Fig. 6.9** Assembly relation transitions and motion macros

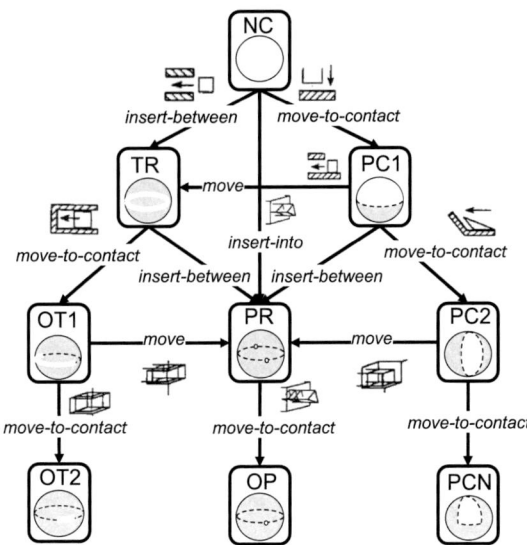

- **move**: This motion macro facilitates the translation of the manipulated object from its initial position to its final position, maintaining its orientation throughout the process.
- **move-to-contact**: This motion macro translates the manipulated object until it makes contact with the surface of the environmental object. Subsequently, it aligns the contact surface of the manipulated object with that of the environmental object.
- **insert-between**: This motion macro initially aligns the manipulated object between a pair of contact surfaces of the environmental object, subsequently translating it between these surfaces.
- **insert-into**: This motion macro aligns the manipulated object along the insertion axis before translating it along the same axis. The primary difference between insert-between and insert-into is that the former requires adjustment in only one direction, whereas the latter necessitates adjustments in two directions.

Figure 6.9 illustrates the motion macros associated with the transition graph of the assembly relationships.

## 6.2.3   Abstract Task Models

We have established thirteen abstract task models corresponding to all possible relationship transitions, as depicted by the arcs in Fig. 6.9. Each task model is represented as a Minsky frame, with each frame containing several slots. These slots store the skill parameters required to perform the transitions and the parameters necessary for those skills. Some parameter slots are pre-filled with constants, while others are left empty and populated with values obtained at runtime by daemons attached to those slots. A task model that is completed by obtaining values at runtime is referred to as an *instantiated task model*, and the process is called as *instantiation*. The followings are representative skill parameter slots:

- **Starting configuration**: This slot stores the configuration of the manipulated object at the onset of the task. The configuration can be derived from the object recognition result or inherited from the final configuration of the object in the task immediately preceding this one.
- **Goal configuration**: This slot encompasses the final configuration of the manipulated object derived from the observation. However, depending on the motion macro such as `move-to-contact`, while this configuration is set as the target, the actual final configuration is determined by the configuration at which contact occurs before or after this.
- **Approach configuration**: For certain skills, it is necessary to adjust the configuration in preparation just before the final complicated motion macros. This slot describes the preparatory configuration for transitioning from a simple `move` motion macro to other complicated macros such as `move-to-contact` or `insert-into` macros for detecting contact force or aligning a hole axis.
- **Approach direction**: This slot contains the direction when approaching to goal configuration. Generally, the direction opposing the normal of the contact surface of environment object obtained from the world model is often used.

Figure 6.10 illustrates the abstract task model representing the transition from NC to PC1, serving as an example of abstract task models. The start and end relationship slots contain the values NC and PC1, respectively. The corresponding skill necessary to achieve this transition is named as the `Place` skill. The motion macro slot encompasses both the `move` motion macro and the `move-to-contact` motion macro. Certain skill parameters, such as the `Approach direction`, are dynamically obtained online by attached daemons. In contrast, others, such as the `Approach distance`, are predefined. The `Approach configuration` is calculated based on the observed `Approach direction`, the observed `Goal configuration`, and the predefined constant `Approach distance`. Furthermore, some parameters, like the `Starting configuration`, are inherited from the previous task model. Other abstract task models also exhibit similar structures.

**Fig. 6.10** `NC-PC1` abstract
task model

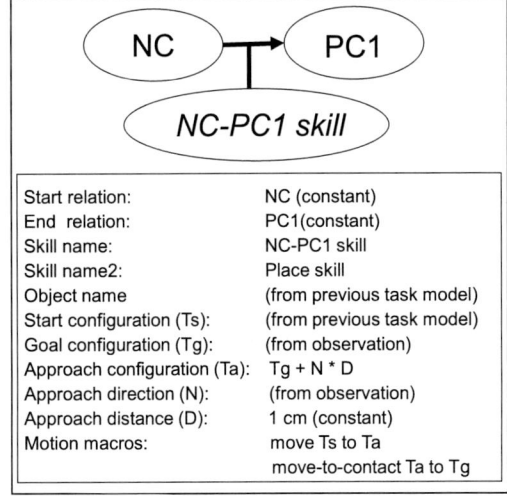

| | |
|---|---|
| Start relation: | NC (constant) |
| End  relation: | PC1(constant) |
| Skill name: | NC-PC1 skill |
| Skill name2: | Place skill |
| Object name | (from previous task model) |
| Start configuration (Ts): | (from previous task model) |
| Goal configuration (Tg): | (from observation) |
| Approach configuration (Ta): | Tg + N * D |
| Approach direction (N): | (from observation) |
| Approach distance (D): | 1 cm (constant) |
| Motion macros: | move Ts to Ta |
| | move-to-contact Ta to Tg |

## 6.3    Runtime System

### 6.3.1   Basic System

#### 6.3.1.1 Task Recognition

The Task Recognition Module (TRM) gathers two distinct types of information through observation:

- **what-to-do**: identifies the transition that occurs, thereby facilitating the retrieval of the appropriate task model.
- **where-to-do**: determines the location where the assembly action should be performed, enabling the collection of suitable skill parameters based on the task model.

Initially, the what-to-do is identified from the transitions of contact states, and the corresponding task model is retrieved. Subsequently, the where-to-do parameters, skill parameters, are collected in accordance with the requirements of the task model. This process is analogous to inserting appropriate operand parameters for a machine instruction, with the what-to-do treated as a single machine instruction. As the number and types of operands differ for each machine instruction, the number and types of skill parameters to be collected vary according to each task (what-to-do).

Task Recognition Module (TRM) determines the what-to-do through human demonstrations. Initially, the TRM compares the two world models obtained from object recognition before and after the demonstration to identify the manipulated object. The assembly relationship is then identified based on the newly generated surface contact between the identified manipulated object and the environmental objects. The TRM then selects the abstract task model corresponding to this transition from the previously established 13 models in the transition graph. Subsequently, the frame of the corresponding task model is instantiated.

Task Recognition Module (TRM) derives the where-to-do parameters based on the descriptions of slots within the task model. The task model retrieved in the previous process contains several empty slots for skill parameters, each associated with a daemon. When the task model is instantiated, the daemons attached to these slots acquire the necessary parameters from the configuration of the objects in the world models. This process aligns with observing the real world through the lens of a task model, as per Minsky's frame theory. Alternatively, this process can be explained using an analogy: the first part identifies the necessary machine instruction, while the second part prepares the required values for the operands of this machine instruction.

**Temporal segmentation** Let's examine the task recognition mechanism using the following example. In this scenario, the system includes two categories of objects—castle and stick— any of which may appear in the scene. See Fig. 6.11. The instructor sequentially retrieves objects from the warehouse and gradually assembles increasingly complex structures.

The Task Recognition Module (TRM) assumes that a human hand enters the scene at the beginning of an assembly action and exits upon its completion. Specifically, the system follows a stop-and-go demonstration approach. In this context, LfO functions as a teaching system, and it is assumed that the instructor sequentially demonstrates actions corresponding to one of 13 predefined tasks. This assumption enables the TRM to segment the continuous image sequence into distinct video clips, each corresponding to a specific task, based on variations in brightness.

The system can detect human actions by analyzing the differences in brightness between consecutive images. In Fig. 6.12, the right side shows an example scene where a human instructor is placing a castle on the table, while the left side displays the consecutive brightness image sequence of that scene. Prior to the human action, the scene consists only of stationary objects (the table), resulting in minimal differences between consecutive images. When the human action occurs, the difference in brightness becomes significant due to the appearance of the human hand within the scene. This disturbance continues until the assembly action is completed and the hand is no longer present in the scene. Once the human hand disappears, the scene again consists only of stationary objects (the castle and the table),

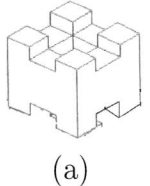

(a)

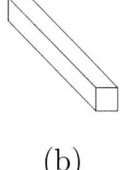

(b)

**Fig. 6.11** Castle and stick

**Fig. 6.12** Image segmentation
based on brightness disturbance

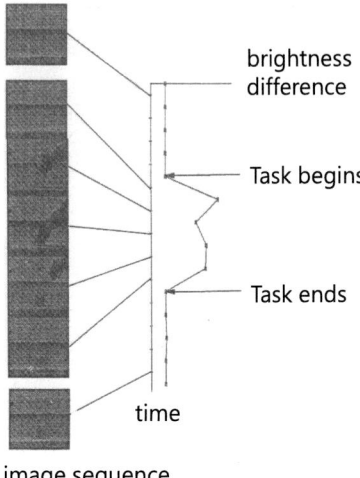

and the difference in brightness becomes minimal. This method provides a robust temporal segmentation of the input scene.

**Object recognition** Range data is obtained for object recognition instead of brightness images. In the current implementation, brightness images are only used to detect the completion of the assembly action, while more reliable range data is employed for object recognition. Upon detecting the completion of an assembly action and after a specified period of time has elapsed, the TRM activates the range sensor and collects range data from the scene.

The manipulated object is recognized based on a difference of range image. The TRM generates a difference range image between the previous step (before the assembly action) and the current step (after the assembly action). Utilizing this difference of range data, the region corresponding to the manipulated object is identified, and the manipulated object is recognized.

The object recognition results are represented in the world model using the Vantage geometric modeler, a geometric modeling tool designed at Carnegie Mellon University (CMU) for this purpose [93]. The world model is incrementally constructed within this geometric modeler based on the recognition results. Initially, the table model is constructed in the world model, and as a manipulated object is recognized during each assembly action, its geometric model is added incrementally to the corresponding positions in the world model. See Fig. 6.13 for reference.

**Task recognition** An assembly relation is obtained by identifying contact pairs between the newly added manipulated object and the previously existing environmental objects in the world model. The system sequentially scans each face of the manipulated object and searches for faces of environmental objects that have the same plane equation, the same

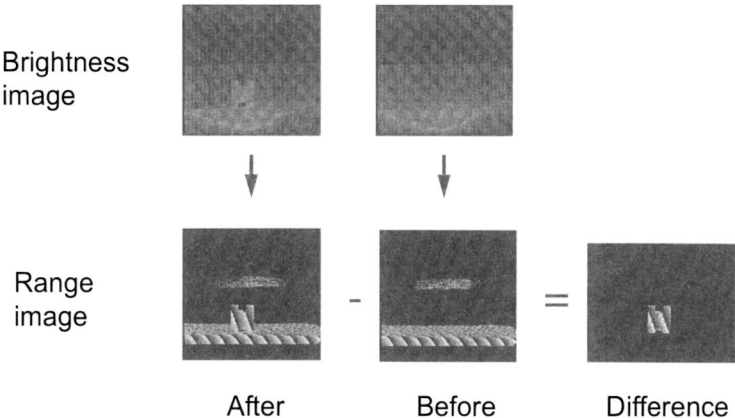

**Fig. 6.13** Depth segmentation based on difference operation

extent, and surface normals pointing in opposite directions. By iteratively examining the linear independence of the generated contact pairs and finding a linearly independent group, the system determines the current contact assembly relationship.

In the example in Fig. 6.14, all four generated contact face pairs, depicted with thick lines, are represented by the same equation, indicating unidirectional contact. Therefore, the current assembly relationship is recognized as PC1. Specifically, after the assembly task is completed, the manipulated object (the castle) is identified to establish the PC1 assembly relationship with the environmental object (the table). Before the assembly task, the castle was not present in the scene, and thus, the assembly relationship between the castle and the table was NC. Consequently, the TRM recognizes that the transition from NC to PC1 occurred during this action. As a result, the NC-PC1 task model corresponding to this transition is retrieved in the transition graph.

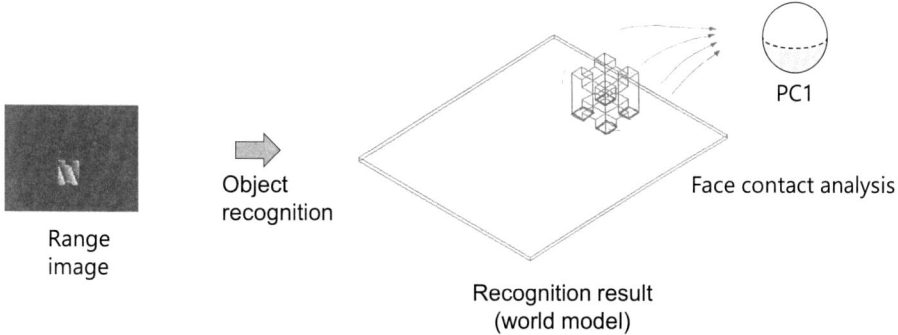

**Fig. 6.14** Task recognition

## 6.3.1.2 Instantiation

**Task model retrieval** Based on the recognition and the pre-provided knowledge, transitions of the assembly relationships for the entire assembly actions that occurred this time are determined as follows:

- PC1-NC: Separate the castle from the warehouse table
- NC-NC: Transport the castle from the initial position
- NC-PC1: Contact the castle with the work table

In the current implementation, the vision system only observes human assembly actions that occur on the table, and it does not observe how new objects are taken from the warehouse table (disassembly actions from the warehouse). This knowledge, as well as the number and position of the objects, is provided to the system in advance. In this example, the system knows that the castle was placed in the PC1 assembly relationship on the warehouse table. Three corresponding task models are instantiated: PC1-NC (Pick), NC-NC (Bring), NC-PC1 (Place).

**Instantiation** In this example, the first two were given to the system as prior knowledge. Therefore, in the following, we will mainly explain the instantiation of the last one, NC-PC1 (Place). The NC-PC1 (place) task model has the following five skill parameters: Start configuration, Goal configuration, Approach configuration, Approach direction and Approach distance. A task model with filled skill parameters from observation and its filling process are referred to as an instantiated task model and instantiation.

The start configuration of this task is the end configuration of the previous task, which is obtained from the previous task model.

The approach direction, along which this task brings the object using the NC-PC1 skill, is defined as the normal direction of the object surface that forms the contact pairs. The approach distance is a predetermined value.

The approach configuration is derived from the approach direction, approach distance and the contact pair configuration. This task initiates the approach action from this specific configuration. Initially, the task executes a move skill from the start configuration to the approach configuration. Subsequently, the task performs the move-to-contact skill from this configuration until a collision occurs. It is important to note that the goal configuration is solely utilized for determining the approach configuration and does not serve as the termination condition. The move-to-contact skill persists until a collision occurs, and the location of this collision defines the task's concluding position.

The task model NC-NC is implemented to teach global motion through the use of waypoints. Collisions between the manipulated object and the environment are not considered by the system. Instead, the demonstrator is assumed to provide intermediate waypoints through a stop-and-go method. By maintaining the hand at each waypoint for a duration, the range sensor is activated to capture the configuration. This process allows for the addition of any number of intermediate waypoints using the NC-NC tasks.

In the current implementation, when the hand holding the object appears on the screen and pauses briefly, the object's configuration is recorded as the end configuration of the NC-NC task. The start configuration is predetermined as the end position of the disassembly operation (PC1-NC) on the warehouse table. Subsequently, the hand resumes its movement to teach the next NC-PC1 task.

The Pick task (PC1-NC) is also implemented as a task model. In the current implementation, the warehouse table cannot be observed due to a limited field of view. Therefore, the PC1-NC task model at the warehouse table is pre-provided knowledge to the system.

### 6.3.1.3 Performance

The system sequentially executes the assembly actions using macro skills in the task models. Figure 6.15 shows the blocks assembled by the robot after observing a human demonstration.

During this demonstration, a recurring issue arose where the rod could not be inserted into the block unless the shape of the hole was adjusted with precision. This problem motivated the development of the error correction system, which will be described in the next section.

## 6.3.2   Error Correction

One of the key advantages of LfO is its ability to recognize the purpose of human demonstrations during observation, enabling the extraction of more precise skill parameters. This awareness facilitates the systematic refinement and adjustment of the observed skill parameters, thereby improving the success rate of robotic actions by fostering a more accurate contextual understanding of demonstrations [10].

As an example of parameter adjustment, consider the assembly step in which a square rod was inserted into the square hole between two castles as part of the assembly process. Figure 6.16a illustrates the state just before this step. One castle was placed atop another, forming a square hole suitable for the insertion of the square rod. However, due to the rough placement of the second castle, the hole's shape was initially unsuitable for proper rod insertion.

**Fig. 6.15** Blocks assembled based on a human demonstration

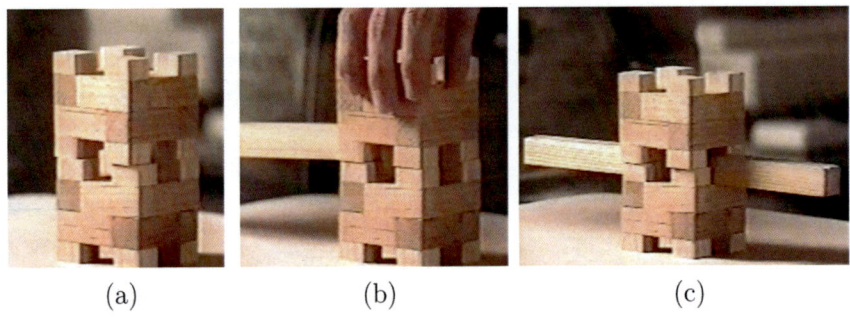

(a)                                    (b)                                    (c)

**Fig. 6.16** Two castles and one rod. **a** Two castles. **b** Before insertion of the rod, the demonstrator adjust the top castle position. **c** The rod inserted into the hole between two castles

As depicted in Fig. 6.16b, immediately before inserting the rod, the demonstrator subtly adjusted the position of the upper castle to ensure alignment, after which the square rod was successfully inserted into the square hole. Figure 6.16c presents the state following the completion of the operation. Notably, the system was not provided with observational data (Fig. 6.16b) regarding the adjustment process; only the final post-operation state (Fig. 6.16c) was given.

In the standard procedure of Learning from Observation (LfO), object recognition is first performed to facilitate the construction of a world model at each assembly step, as illustrated in Fig. 6.17. Figure 6.17a represents the outcome of the assembly operation in which one castle was placed on top of another. Figure 6.17b shows the result of the square rod insertion operation. During this step, only the position of the manipulated object is updated using difference of range images. Due to threshold settings, the upper castle—despite undergoing slight rotation—is not classified as a manipulated object; consequently, its position remains consistent with previous recognition results. In contrast, the square

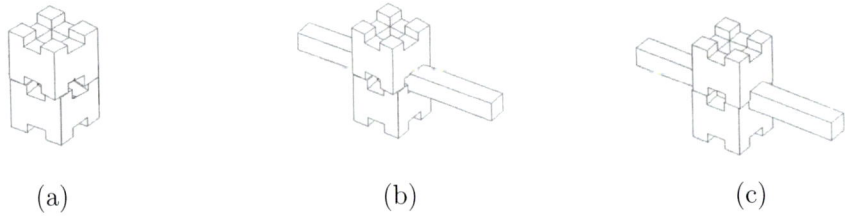

(a)                                    (b)                                    (c)

**Fig. 6.17** World models. **a** Pre-operation world model. One castle was placed on top of another. Due to its rough placement, the shape of the hole became unsuitable for insertion. **b** Post-operation world model (without any adjustment). The square rod was inserted. Since only the position of the rod was recorded in the world model as the manipulated object, the rod ended up penetrating the surface of the surrounding hole. **c** Post-world model (with adjustment). Based on the recognition results of NC-PR, the world model was updated with a corrected position for the upper castle

rod, having undergone substantial movement, is detected via difference of range images, recognized as a manipulated object, and incorporated into the world model. As a result, in the post-operation world model depicted in Fig. 6.16b, the rod appears to have penetrated the surrounding hole.

To explore the newly established surface contact relationship between the manipulated object and the environmental objects, as represented in the world model shown in Fig. 6.16b, pairs of surfaces with opposing normal directions are analyzed between these objects. By applying a relatively large threshold to accommodate directional errors in the contact relationship exploration, it is observed that, following the operation, the square rod establishes a PR relationship with the square hole in the castle. Consequently, this observation confirms that a transition from NC to PR has occurred as a result of this assembly operation. The motion macro to execute is identified as move-insert.

If the move-insert motion macro is executed using the current world model without modifications, an operational error is inevitable. Figure 6.18 presents an example in which the operation was performed based on the original world model parameters without any adjustments. In Fig. 6.18a, the upper castle is placed on top of the lower castle according to the world model parameters. The square rod is tried to be inserted without any adjustments, as shown in Fig. 6.18b. Due to misalignment of the square hole, the square rod cannot be inserted and instead pushes against both castles in Fig. 6.18c. After the move-insert macro is completed, the square rod is released. Since it was not successfully inserted into the hole, it falls, as shown in Fig. 6.18d.

The adjustment of object locations in the world model to successfully execute the move-insert operation can be achieved by fine-tuning the coefficients so that the error between equations in each surface contact relationship remains below a predefined threshold. In general, solving this problem is extremely challenging due to two main factors: nonlinearity and cyclic dependencies. Fortunately, human demonstrations provide valuable insights for overcoming these difficulties.

 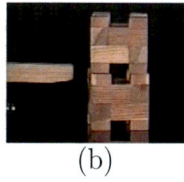  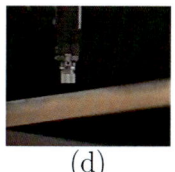

(a)                    (b)                    (c)                    (d)

**Fig. 6.18** Execution based on the unmodified world model parameters. **a** The upper castle is placed on top of the lower castle according the object location in the unadjusted world model. **b** The square rod is tried to be inserted without any adjustments. **c** Due to misalignment of the square hole, the square rod cannot be inserted and instead pushes against both castles. **d** After the move-insert motion macro is completed, the square rod is released. Since it was not successfully inserted into hole, it falls

(a)                    (b)                    (c)                    (d)                    (e)

**Fig. 6.19**  Insertion of the rod into the hole after world model adjustment. **a** The upper castle is placed on top of the lower castle. **b** Due to inconsistencies in the world model parameters, the upper castle is lifted again to align the hole's shape. **c** The necessary rotation is applied to the upper castle, and it is placed back onto the lower castle. **d** The insertion is then attempted. **e** Successful insertion

Regarding nonlinearity, since the demonstration provides approximate object positions, the correction equations can be linearized using Taylor expansions of the nonlinear equations around their current values, which include sine and cosine functions.

As for cyclic dependencies, since the order of object manipulation is known from the demonstration, the correction amount for each object can be sequentially calculated, beginning with the most recently manipulated object. This approach allows for determining small displacements that ensure the overall equation error remains within a certain threshold. Figure 6.17c presents the updated world model based on this adjustment. For further details, refer to [10].

Figure 6.19 illustrates an example in which the `move-into` macro was successfully executed following an adjustment to the upper castle's position, based on the parameters shown in Fig. 6.17c. Figure 6.19b and c depict the adjustment process: in Fig. 6.19b, the upper castle is picked up for adjustment, and in Fig. 6.19c, it is placed again using the adjusted parameters. After completing the adjustment process, the square rod is successfully inserted, as shown in Fig. 6.19d and e. Refer to the YouTube video[2] for details on the system's movements.

### 6.3.3  Mechanical Components

#### 6.3.3.1 Relationship in Mechanical Components

To apply LfO to the assembly of mechanical components, it is necessary to introduce relationships specific to mechanical parts in addition to the polyhedral relationships observed so far. Commonly encountered relationships in mechanical components include the pairing of cylindrical shafts with their bearings, spherical ball bearings with their housings, bolts with threaded holes, and the alignment of gears. These consist of relationships between curved surfaces as well as connections unique to mechanical components.

---

[2] https://youtu.be/mZMqcuH1bxs.

**Curved surfaces**  Ikeuchi et al. conducted an in-depth examination of surface contact relationships involving curved surfaces, which play an important role in the assembly of mechanical components [177]. Notable examples requiring the consideration of such surface contact relationships include the placement of a cylindrical shaft within its bearing and a spherical ball bearing within its housing. They demonstrated that the relationships for polyhedra described in the previous subsection can be extended to handle curved surfaces through the addition of multiple branches to the existing transition graph.

The mathematical representation of surface contact is conveyed through the distribution of normal vectors at the contact points. Notably, the governing equations for these relationships exhibit no fundamental differences between planar and curved surfaces because the normal vector pairs at the contact points are defined based on the tangential planes, for both planar and curved surfaces.

Within Vantage, curved surfaces are represented as a collection of planar patches—for example, a cylinder is modeled as a polygonal column. These planar patches retain the characteristics of curved surfaces while integrating the equations that define them. Consequently, analyzing surface contact relationships involves examining the distribution of normal vectors derived from the equations of adjacent patches belonging to both the manipulated object and the environmental object. This methodology facilitates the determination of contact relationships for curved surfaces in a manner consistent with those applied to planar surfaces.

Curved surfaces, however, introduce additional constraints on the permissible direction of motion due to the second-order differentiability of curved surfaces. For example, when a cylindrical shaft is placed on a semi-circular bearing, as illustrated in Fig. 6.20a, the state where the cylindrical shaft rests within the open bearing is defined as OT1 state. Within this configuration, allowable movements include axial displacement or detachment perpendicular to the cylinder's axis. In either case, the transition proceeds directly to NC state without passing through TR state due to its second-order characteristics. To accommodate this transition, an NC-OT1 branch was added to the transition graph, and the corresponding skill was named as `cylinder-to-contact`.

For implementation of this skill, there are two approaches to achieving transitions: one through motion perpendicular to the axis of the bearing and another along the axis. Following the same principle used for transitions between polyhedral surfaces, the rule prioritizes perpendicular motion, which is adopted as the movement direction for the `cylinder-to-contact` skill. First, based on visual feedback and other information similar to the `insert-between` motion macro, the axial alignment between the bearing and the cylindrical shaft is adjusted. After this adjustment, the cylindrical shaft is guided onto the bearing using the same control as the `move-to-contact` motion macro.

Similarly, in the case of a sphere and its housing in Fig. 6.20b, an NC-OP branch was introduced based on the same reasoning, with the required skill designated as `sphere-to-contact`. For implementation, control is first applied similarly to `insert-into`, followed by control similar to `move-to-contact`.

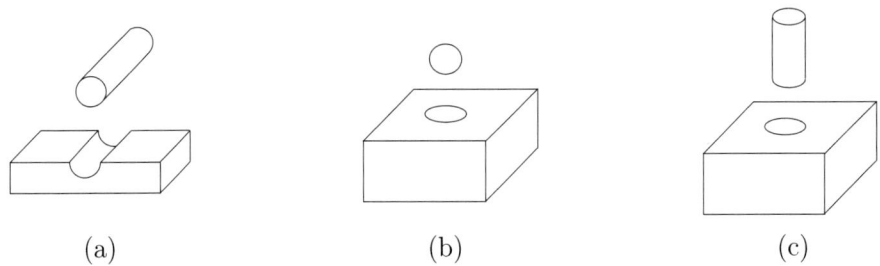

(a)                               (b)                               (c)

**Fig. 6.20** Curved surface fittings commonly encountered in machine assembly. **a** Positioning a cylindrical shaft into a housing. `cylinder-to-contact`. **b** Aligning a spherical ball bearing into its housing. `sphere-to-contact`. **c** Inserting a cylindrical shaft into a hole. `insert-into`

For scenarios involving the insertion of a circular shaft into a circular hole, as depicted in Fig. 6.20c figure, the transition remains consistent with NC-PR, resulting in `insert-into`, identical to the process observed in polyhedral objects. Consequently, no additional modifications to the framework are required.

**Bolts and Gears** Miura et al. worked on handling relationships specific to machine parts, such as bolts/nuts and gears [178]. In Vantage, nut holes with internal threads are represented as cylindrical surfaces, and bolts with external threads as cylindrical rods. Attributes like internal/external threads were attached to corresponding cylindrical surfaces. See Fig. 6.21a.

Transition branches for machine parts were developed in parallel with existing transitions to ensure they are tailored to their respective attributes, as illustrated in Fig. 6.25. A bolt operation begins with the insertion of the bolt into a nut hole, modeled as the insertion of a cylindrical rod into a cylindrical hole, as described by Ikeuchi [177]. This process activates a transition of NC-PR, utilizing the `insert-into`. Once the bolt-tightening operation is detected – where the external and internal threaded surfaces are fully in contact – the tran-

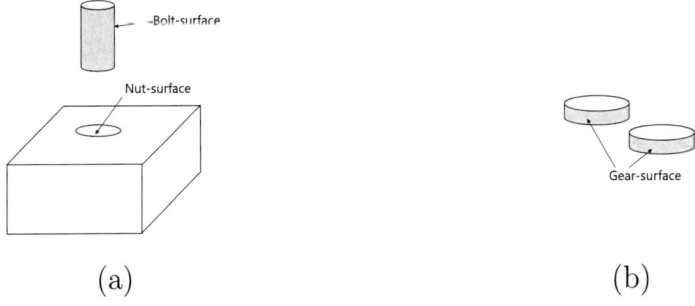

(a)                                              (b)

**Fig. 6.21** Relationships of mechanical components. **a** Bolt-tightening. **b** Gear-mating

sition from PR-OP employs the `bolt-tightening` instead of `move-to-contact`, based on the attributes of the paired surfaces. To accommodate this, an additional branch is incorporated into the PR-OP transition.

A dedicated branch for the `gear-mating` skill is incorporated specifically for gears. In Vantage, gears are represented as cylindrical surfaces, as illustrated in Fig. 6.21b. When contact occurs between convex cylindrical surfaces, it triggers the NC-PC1 transition. Given that these cylindrical surfaces are identified as gear surfaces, an additional branch is introduced to activate `gear-mating` instead of `move-to-contact`, ensuring the transition aligns with the specific attributes of the surfaces

### 6.3.3.2 Implementation of `bolt-Tightening` and `gear-mating`

This subsection discusses the implementation of `bolt-tightening` and `gear-mating` skills, both of which rely heavily on visual feedback. In the case of `bolt-tightening`, it is assumed that the bolt has already been inserted into the nut hole. Similarly, for `gear-mating`, it is presumed that the axis of the target gear has been positioned within the hole. Although these operations inherently involve rotational movements, this discussion adopts a translational motion framework due to the small-scale nature of the movements, ensuring simplicity.

**Bolt-tightening** The implementation of `bolt-tightening` largely depends on the hardware used. Here, we consider the most common scenario, where a gripper holds the screwdriver to perform the operation as shown in Fig. 6.23c. `Bolt-tightening` can be divided into two distinct sub-stages: first, aligning the screwdriver with the slot of the bolt; and second, rotating the screwdriver until sufficient resistance is generated. Since the second action is relatively straightforward, this discussion will focus primarily on the first action, which involves aligning the screwdriver with the slot on the bolt head.

The alignment of the screwdriver can be conceptualized as the `insert-between` motion macro, as illustrated in Fig. 6.22a. Here, S-axis represents the motion direction, which corresponds to the approach direction of the screwdriver. The axes T and U, defined as orthogonal to S-axis, are introduced as shown in Fig. 6.22b.

In the T-axis direction, the state transitions from maintenance when in free space to constraint when the screwdriver head is inserted into the slot of the bolt head. Accordingly, rule B9 is applied in this direction.

```
B9: if BeforeTransition:
        if |T - feature-t| > delta-gap, then penalty
    else:
        if F-t > delta-collision, then penalty
```

B9 specifies that, prior to the transition, feature points on the environmental object must be aligned with those on the manipulated object along the T-axis. To achieve this alignment,

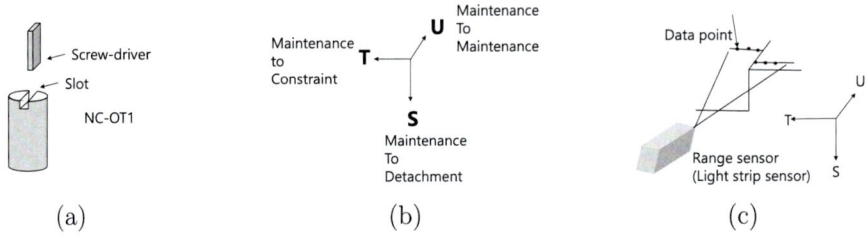

(a)                                 (b)                                 (c)

**Fig. 6.22** `insert-between` motion macro in `bolt-tightening` skill. **a** Slot and screw driver. **b** Coordinate system and corresponding transition of directional contact states **c** Advantageous sensing direction with respect to the direction of contact transition (T-axis direction)

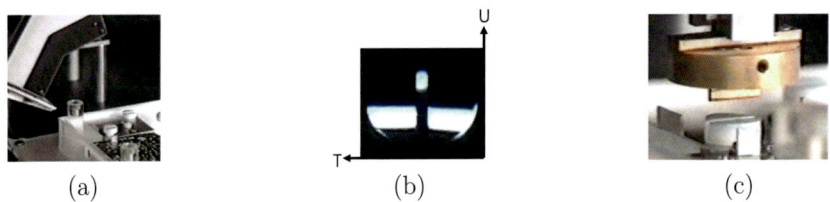

(a)                                 (b)                                 (c)

**Fig. 6.23** Implementation of `bolt-tightening` skill. **a** Placement of a range sensor along the S-U plane. **b** Obtained range image along the T direction. **c** Execution of `bolt-tightening` skill

the T-coordinate of the environment feature with respect to the robot coordinate should be extracted beforehand. For this purpose, it is generally advantageous to conduct sensing in a direction orthogonal to the axis along which the transition from maintenance to constraint occurs—specifically, the T-axis. More precisely, sensing should be performed in the direction that lies within the plane defined by the S-axis and U-axis (see Fig. 6.22b and c).

Figure 6.23 illustrates the implementation of the `bolt-tightening` by Miura et al. First, by positioning a range sensor along the U-S plane relative to the grove of the bolt and projecting light, it is possible to obtain range information in the T direction. Based on this information, the alignment of the screwdriver is performed, followed by the execution of the rotation operation. For further details, refer to [48].

**Gear-mating** A similar approach can be applied to gear mating. In Fig. 6.24a, the motion direction is defined as the axis perpendicular to the plane of this page. The T-axis, which is orthogonal to the motion direction, transitions from Maintenance to Constraint. To properly align the manipulated object along this axis, it is beneficial to gather positional data by positioning a range sensor along the S-U plane, which is perpendicular to the transition direction, T, as shown in Fig. 6.24a. Figure 6.24b illustrates the sensor alignment in relation to the environment and manipulation objects.

Figure 6.24c illustrates the pre-adjustment and post-adjustment range images obtained. Based on the correction values derived from these images, a gear mating was conducted and

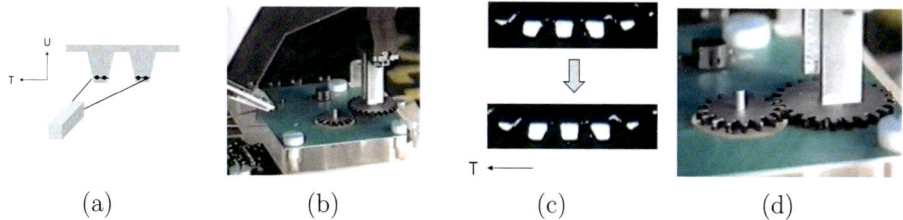

(a)                    (b)                    (c)                    (d)

**Fig. 6.24** `Gear-mating` skill. **a** Sensor planning. **b** Range sensor placement. **c** Alignment adjustment. **d** `gear-mating` skill execution

`gear-mating` successfully completed its operation, as shown in Fig. 6.24d. Examples of these implemented skills can be found in a YouTube video, *Learning-from-observation: Task-oriented generation of sensing strategy.*[3]

### 6.3.3.3 Abstract Task Models for Mechanical Components

The abstract task models for handling mechanical components is illustrated in Fig. 6.25. In addition to the fundamental case of polyhedral objects, it includes the following additional task models:

- **NC-OT1**: invokes the `cylinder-to-contact` skill for handling cylindrical surfaces.
- **NC-OP**: invokes the `sphere-to-contact` skill for handling spherical surfaces.
- **PR-OP**: invokes the `bolt-tightening` skill for bolt/nut pairs.
- **NC-PC1**: invokes the `gear-mating` skill for gear surfaces.

### 6.3.3.4 Runtime System

This system maintains a fundamental schema aligned with previous implementations, employing the stop-and-go method to acquire range images based on the presence or absence of brightness fluctuations. However, it introduces noteworthy advancements. In addition to capturing pre-operation and post-operation range images, the system obtains a during-operation range image for the instruction of the object's grasping position. For example, in the `Place` task depicted in Fig. 6.26, after placing the manipulated object in its designated location, the hand is briefly paused. This brief pause activates the range sensor, enabling the recording of a range image encompassing both the object and the hand grasping it. Subsequently, releasing the hand facilitates the acquisition of the post-operation range image.

---

[3] https://youtu.be/Rn-rT4iq30Q.

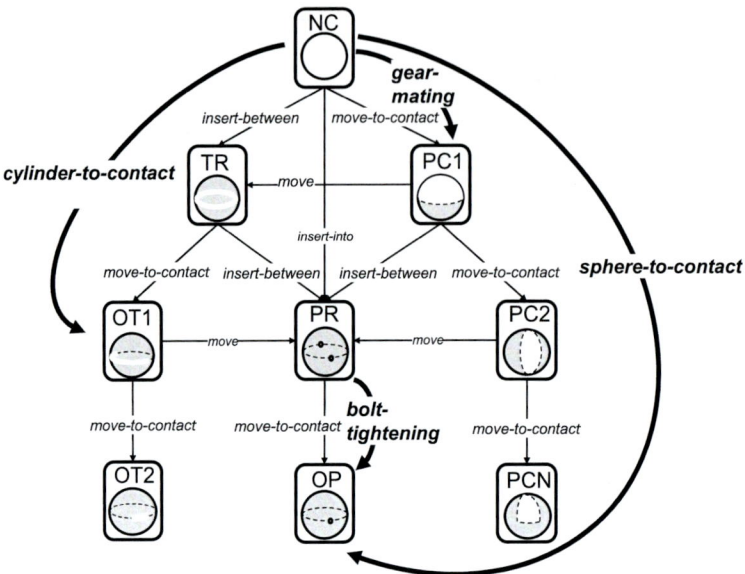

**Fig. 6.25** Abstract task models for handling mechanical components. The basic set is the same as that for polyhedral objects, but several additional models have been incorporated

Figure 6.26a illustrates the initial scene as part of the preparation phase. The range image of this scene is obtained using a light-stripe sensor, as depicted in Fig. 6.26d. Subsequently, an object is introduced into the field of view from outside. This action generates brightness vibrations, and by keeping the hand stationary and allowing the brightness to stabilize, it serves as a trigger to invoke the range sensor. This enables the acquisition of the during-operation range image, capturing both the hand and the object, as shown in Fig. 6.26b.

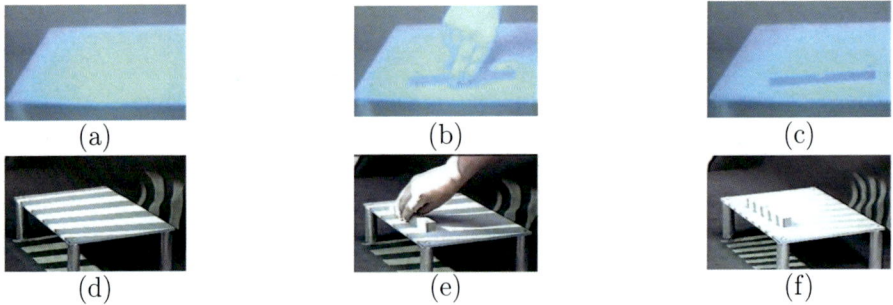

**Fig. 6.26** Input scene. **a** Pre-operation scene. **b** During-operation scene. **c** Post-operation scene. **d** Pre-operation range-data acquisition. **e** During-operation range-data acquisition. **f** Post-operation range-data acquisition

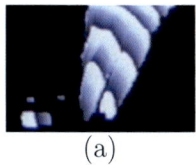
(a)

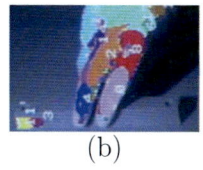
(b)

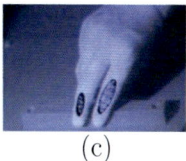
(c)

**Fig. 6.27** Identification of grasping position. **a** Extracted hand region. **b** Segmentation results. **c** Results of superquadric surface fitting

Following this, the hand is moved out of the field of view, causing a change in brightness and facilitating the acquisition of the post-operation range image, which captures only the object on the table, as shown again in Fig. 6.26f.

The range image of the hand is derived by calculating the difference between the during-operation and post-operation range images. Figure 6.27a presents the resultant range image of the hand. By applying a superquadric surface fitting to the segmented range image, the grasping position is identified, as demonstrated in Fig. 6.27c.

The range image of the manipulated object is obtained by calculating the difference between the post-operation and pre-operation range images. The resulting range image is identified as a bar, and its 3D model is integrated into the world model in Vantage. The contact relationship between the manipulated object and the environmental object (the table) is analyzed within the world model and classified as a PC relationship. Consequently, the transition is identified as NC-PC, and the Place skill is deemed necessary, leading to its execution, as illustrated in Fig. 6.28a. It is noteworthy that the grasping position defined in Fig. 6.27c is employed.

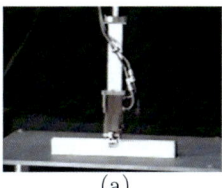
(a)

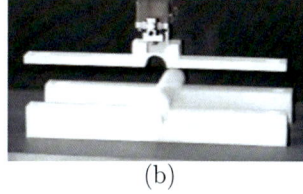
(b)

**Fig. 6.28** Execution examples. **a** The surface contact relationship transition is recognized as NC-PC, and the Place skill is executed. **b** The surface contact relationship transition is recognized as NC-OT1, and, due to the attribute of contact pairs is the cylindrical surface, cylinder-to-contact skill is executed

Figure 6.28b illustrates the process of placing the upper bearing onto a cylindrical shaft after several steps of this procedure. The transition in surface contact relationships is classified as NC-OT1, and based on the cylindrical properties of the surface, the cylinder-to-contact skill is executed.

Figure 6.29 illustrates the demonstration scenes related to bolt-tightening. Initially, the demonstration is recognized as an NC-PR transition. Subsequently, it is identified as a PR-OP via bolt-tightening skill. Note that during the bolt-tightening by the demonstrator, the grasping position for the other hand, intended for support, is also taught.

Based on the demonstration, the NC-PR skill is executed first, involving the insertion of the bolt into the hole, as illustrated in Fig. 6.30a. Subsequently, the bolt is tightened using the bolt-tightening skill with a screwdriver. During this procedure, the other hand grips the object for support, as shown in Fig. 6.30b and c.

Figure 6.31 illustrates the final assembly, successfully completed based on the demonstrations. A video showcasing the operational dynamics of the system,

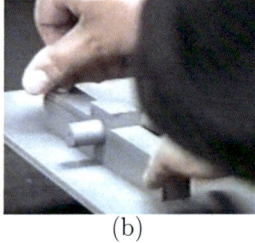

(a)                                              (b)

**Fig. 6.29** Demonstration images before and during bolt-tightening. **a** Bolt insertion (NC-PR). **b** Bolt-tightening (bolt-tightening

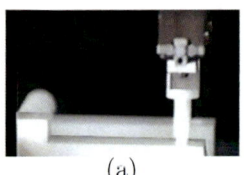
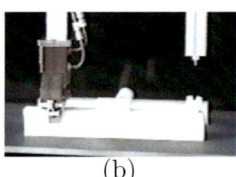
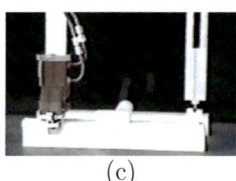

(a)                              (b)                              (c)

**Fig. 6.30** Bolt-tightening execution. **a** Screw insertion. **b** Approach of the screwdriver for bolt-tightening. **c** Bolt-tightening operation

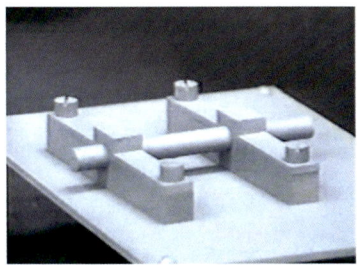  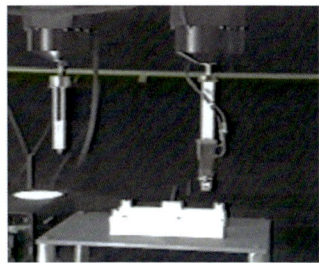

**Fig. 6.31** Final assembly successfully assembled

*Learning-from-observation: Planar, curved, and mechanical contacts* has been recorded and uploaded to YouTube, providing a visual representation of its functionality.[4] Those interested are encouraged to consult the video for further insights.

---

[4] https://youtu.be/gsWyxdMvMIc.

# Knot Tying World

**7**

This chapter investigates a knot-tying system as a case study for handling flexible objects, applying the LfO frameworks introduced in the previous section. The systems discussed in earlier chapters primarily addressed rigid bodies, such as polyhedrons, by associating requisite actions with contact-state transitions. Building on these concepts, this chapter extends the focus to flexible objects, specifically examining a knot-tying system. The system leverages representations of knot types and their transitions to enable the generation of knot-tying actions.

In this system, the objective is to tie knots using a string placed on a table. The knot-tying system observes the string on flat surfaces, such as tables, as demonstrated by humans, to create various knots. Although the ultimate goal is to tie knots in mid-air, this study focuses on analyzing knot configurations by circumventing the complexities of grasping and detecting strings in mid-air. Therefore, demonstrations involve tying a string on a table, and the resulting two-dimensional projection images of knots are analyzed based on transitions of the representations from Knot theory. The system then infers the human's tying actions based on these transitions. Subsequently, the robot aims to replicate similar knots on the same table based on these analyses.

The primary representation in this system is the P-data, derived from Knot theory [11, 179, 180], which corresponds to surface contact in the Kuhn-Tucker theory. Consequently, the system workflow is as follows:

- **Pre-processing of range images**: The vision system captures a series of range data of the knots created by the demonstrator and extracts the state of the string from them. The range data is utilized to extract not only the current shape of the string but also the relative positioning (up or down) of string parts at the intersections.

K. Ikeuchi et al., *Learning-from-Observation 2.0*, Synthesis Lectures on Computer Vision, https://doi.org/10.1007/978-3-032-03445-8_7

- **State recognition**: The extracted shape and relative positioning of the string are converted into a representation known as P-data. P-data maintains a one-to-one correspondence with the current topology of the string. Specifically, when a projection of the string is given, a unique P-data is obtained; conversely, the original projection can be reconstructed from the P-data. This demonstrates that the knot state of the string can be adequately and sufficiently described using P-data. The extraction of P-data is analogous to the extraction of surface contact relationships from object recognition, as discussed in the previous chapter.
- **Task recognition**: The system identifies the appropriate motion skill based on the knot state transition. Knot theory posits that all knot transitions can be realized by a combination of four moves: Reidemeister I, II, III, and cross moves. To identify the appropriate motion skill, we first examine the P-data transition in a pair of P-data before and after an action. Subsequently, we determine which Reidemeister or cross move occurred based on the transition. This process corresponds to the extraction of motion skills from contact relationship transitions discussed in the previous chapters.
- **Skill parameter determination**: Skill parameters, such as the direction of twisting or the location of movement required to execute each motion skill, are derived from P-data following the identification of the necessary motion skill. This determination is made by analyzing specific attributes within designated portions of the P-data for each motion skill. This approach corresponds to the process of acquiring affordance parameters for each task, as described in the previous chapters.
- **Execution**: Sequentially invoke agents capable of executing the motion skills to recreate the same knot.

In the following sections, we will discuss P-data and Reidemeister moves, as well as their relationships. Subsequently, we will explain the implementation of the system.

## 7.1    P-Data Representation

This section introduces P-data as a representation of knots in an open-curve string. Traditional Knot Theory only handles closed-curve knots forming continuous loops. However, applying this framework directly can be impractical for our purposes. Therefore, we redefine the string as an open curve with two fixed endpoints. By assuming that these endpoints remain stationary before and after any action, the open curve can effectively be treated as a closed curve, allowing the application of Knot Theory. For scenarios where the endpoints are not fixed and may move, additional theoretical frameworks will be developed to address such cases.

The essence of Knot theory lies in the ability to reconstruct the global topology from the local relationships at the intersections in the projection image. By projecting an open curve in three-dimensional space onto a plane from an appropriate viewpoint, we obtain a two-dimensional open curve. P-data describes the global topology based on the local analysis of the projection.

For Knot theory to be applicable to a two-dimensional project, we will choose the general viewpoint that satisfies the following conditions:

• Each intersection is a point intersection rather than a line intersection.
• No three parts of the string intersect at the same point.
• None of the ends lie on any part of the string.
• The relative positioning of the string parts at each intersection can be discernible.

From a general view point, each intersection consists of two points: one located on a nearby segment of the string and the other on a distant segment. These are referred to as the upper intersection point and the lower intersection point, respectively. Henceforth, the term "point" will specifically denote an intersection point, while the term "segment" will refer to a portion of the string that is enclosed either by two points or by one point and the string's endpoint.

P-data is obtained from the local analysis of the spatial (up/down and left/right) relationships between the points and nearby segments in the projection. The resulting P-data is known to have the following properties:

• P-data depends on topology and not on local deformations.
• The conversion from the knot projection to P-data is reversible [180].

The aforementioned properties of P-data are convenient for describing knots. The first property ensures that P-data does not change due to the local movement of intersections or the local transformation of string shapes. For instance, as shown in Fig. 7.1, although the projections of two knots may differ, they are essentially equivalent in terms of knot formation,

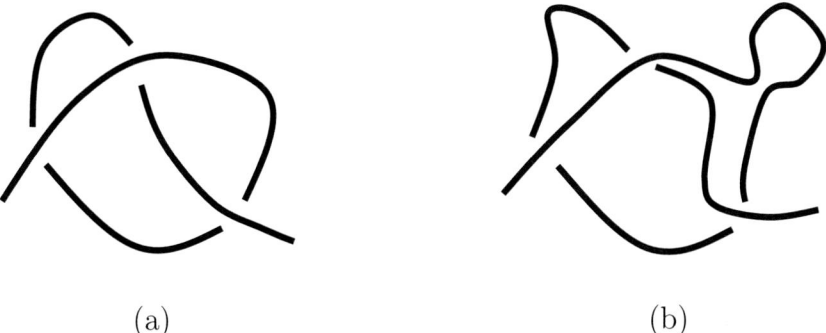

(a)                                                                                  (b)

**Fig. 7.1** Two equivalent knots in the P data representation

and both have the same P-data. The second property guarantees that the projection and P-data have a one-to-one correspondence; thus, the P-data obtained from a projection is a unique representation.

### 7.1.1   P-Data Generation

P-data consists of the number of the point, the number of the counterpoint, as well as spatial characteristics of each point. P-data is generated from a knot projection in the following four steps: starting point selection, point enumeration, attribute assignment, and compilation.

**Starting point selection**: In P-data, we first select one of the two end points of the string as the *starting point*, and the other end point is referred to as the *terminal point*.

**Point enumeration**: Beginning from the starting point, we progress along the knot projection until we reach the terminal point. The direction of enumeration along the string is referred to as the *enumeration direction*. When encountering a point (an upper or lower intersection point), we assign a number to it based on the order of occurrence. As shown in Fig. 7.2, numbers are assigned sequentially as we encounter intersection points from the left end of the string. Consequently, each intersection has two numbers: one for the first encounter and another for the second encounter.

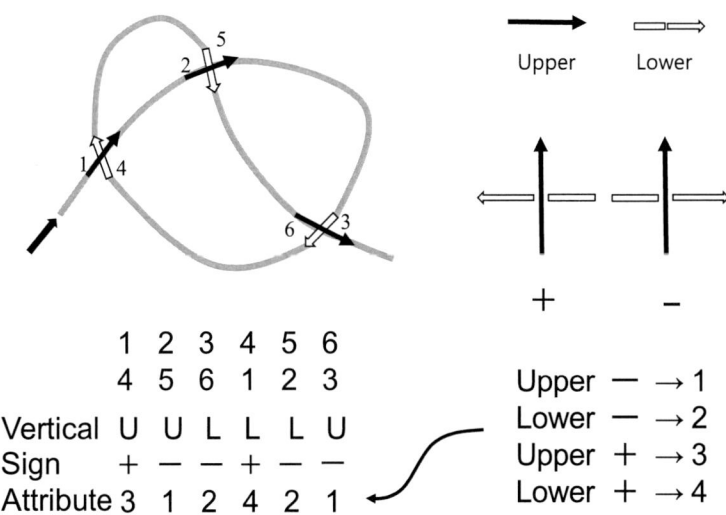

**Fig. 7.2**  Knot projection and P-data

**Characteristics assignment**: For each point, we determine the counterpart, the vertical position sign (upper or lower), and the direction sign (plus or minus). The counterpart of the point is first identified based on the results of the previous step. Then, the designation of *upper* or *lower* is assigned based on whether the point is on the string portion that passes above or below the other string portion. Additionally, the direction sign is assigned to indicate whether the string crossing direction is to the left or right. If the lower portion passes from the right to the left with respect to the enumeration direction of the upper portion, it is assigned a *plus* sign; otherwise, it is assigned a *minus* sign.

**Attribute compilation**: The combination of these four signs (upper/lower and plus/minus) is assigned to each point as its attribute as follows:

- upper and minus, then one
- lower and minus, then two
- upper and plus, then three
- lower and plus, then four

P-data is the combination of the current point number, the counterpoint number, and its attribute.

From the aforementioned process, the P-data for the knot in Fig. 7.2 is obtained as follows

$$
\begin{matrix}
1 & 2 & 3 & 4 & 5 & 6 \\
4 & 5 & 6 & 1 & 2 & 3 \\
3 & 1 & 2 & 4 & 2 & 1
\end{matrix}
\tag{7.1}
$$

For the sake of future discussions, we define some terms and functions.

- The $i$-th point is defined as the $i$-th encountered point during the process of converting the knot projection to P-data. Similarly, the $i$-th segment is the $i$-th encountered segment during the process.
- We denote the current step number of knot-tying actions as $t$, and the corresponding current P-data at that step as $P_t$. Analogous to the assembly of polyhedra, we assume that the demonstrator performs each deformation action of the string sequentially, pausing after each action to present the result to the system. The number of current action in this process is denoted as $t$. When the context is clear, we will abbreviate it as $P$.
- $n(P)$ represents the number of columns in $P$. This corresponds to the total number of points along the string, excluding the starting and end points, as well as twice the number of apparent intersections in the projection. Note that each intersection consists of points: one above and one below.
- $\sigma(i|P)$ represents the number of the other point at the $i$-th point. If $i$ corresponds to an upper point number, then $\sigma(i|P)$ corresponds to the lower point number, and vice versa.

Thus, if $\sigma(i|P) = j$, then $\sigma(j|P) = i$ must also hold, ensuring that the commutative property is consistently satisfied. When the context is clear, we simplify the notation to $\sigma(i)$.

- $attr(i|P)$ represents the attribute of the $i$-th point, which can be one of the following: (1) Upper and minus; (2) Lower and minus; (3) Upper and plus; (4) Lower and plus. When the context is clear, we simplify the notation to $attr(i)$.

For example, in the above P-data, $n(P) = 6$, $\sigma(2|P) = 5$, $\sigma(5|P) = 2$, and $attr(3|P) = 2$.

## 7.2    P-Data Transition and Motion Skills

### 7.2.1    Motion Skills

We define four motion skills to handle an open curve, consisting of three types of Reidemeister moves from Knot theory and one Cross move. The three Reidemeister moves are fundamental actions as defined in the original Knot theory. However, since the original Knot theory pertains to closed curves, we have introduced the Cross move to adapt it for open curves.

According to Knot theory, two knots are considered equivalent if one can transition to the other without cutting the string. Reidemeister demonstrated that any equivalent knot transition can be achieved by repeatedly applying three types of moves, known as Reidemeister moves, a finite number of times. Figure 7.3 shows examples of Reidemeister moves.

- **Reidemeister I**: This move involves adding or removing one single intersection by creating or destroying a simple loop. See Fig. 7.3a.
- **Reidemeister II**: This move entails adding or removing two intersections by having one segment of the string cross over another segment. See Fig. 7.3b.
- **Reidemeister III**: This move involves passing a segment of the string through an intersection, without changing the number of intersections. See Fig. 7.3c.
- **Cross**: In Knot theory, a string is considered a simple closed curve, meaning a tangled loop without end points. However, in practical applications, a string has two end points, allowing for the movement of these end points without cutting the string itself. Such moves are not accounted for in the Reidemeister moves derived from Knot theory. Consequently, we define the Cross move as an additional motion skill. This skill involves adding or removing an intersection by having the end point of the string cross a segment. See Fig. 7.3d.

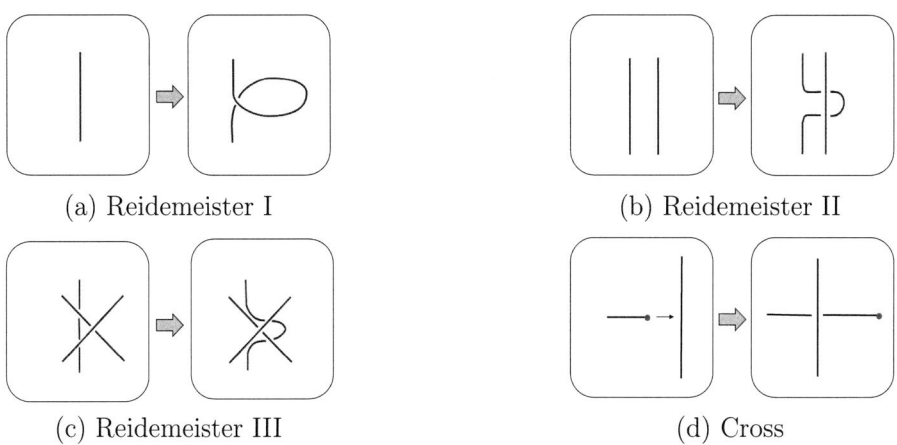

(a) Reidemeister I                    (b) Reidemeister II

(c) Reidemeister III                    (d) Cross

**Fig. 7.3** Reidemeister and cross moves

By repeatedly applying these four motion skills, it is possible to create typical knots such as the *overhand knot*, *figure-eight knot*, *bowline knot*, *harness hitch*, *bow tie*, *single loop bow*, *two half hitch*, and *taut-line hitch* as shown in [11].

### 7.2.2   Assumption for Transition Analysis

We will describe the method to estimate the motion skills that caused the observed transitions in P-data. Here, $P_t$ denotes the P-data obtained from the knot projection at time $t$. Without loss of generality, we can assume that the knot projection before the transition, $P_{t-1}$, has fewer intersections than the knot projection after the transition, $P_t$. This corresponds to the process of gradually creating a more complex knot from a single piece of string. That is to say,

$$n(P_t) \geq n(P_{t-1}). \tag{7.2}$$

This analysis also assumes that

- The demonstrator shows the results to the system following each deformation action,
- Each deformation action is either a Reidemeister-move skill or a cross-move skill,
- Each deformation action is applied exclusively to one segment.

Since P-data depends on the choice of the starting point among two end points, we assume that the same point is used as the starting point for transition analysis before and after each deformation action.

For simplicity of explanation, the following discussion uses the direction in which the number of points increases. Therefore, the description is based on the P-data after the operation has taken place. In the case of reverse operations, it is sufficient to analyze the P-data before the operation occurs.

### 7.2.3  Transition Analysis

#### 7.2.3.1  Reidemeister I Moves and Corresponding P-Data Transitions

When Reidemeister I is applied to a string, one intersection is added to the knot projection. Since two new points (upper and lower) at the intersection are generated, the following conditions must be satisfied.

$$n(P_t) = n(P_{t-1}) + 2.$$

Conversely, if the number of columns of P-data increases by two from $t-1$ to $t$, it indicates the possible application of a Reidemeister I move.

Figure 7.4 shows an example of the transition of P-data with a Reidemeister I move applied to segment 3. Two new intersection points, 3 and 4 are generated. As a result, initially, the number of points, $n(P_{t-1})$ is 6, but it undergoes an increase, reaching $n(P_t) = 8$.

Assume a Reidemeister I move is applied to the $i$-th segment. The added two points will have consecutive numbers and an upper/lower relation with each other. Thus, at the time $t$,

$$\sigma(i) = i + 1,$$
$$\sigma(i + 1) = i,$$
$$|attr(i) - attr(i + 1)| = 1.$$

In the example shown in Fig. 7.4, the consecutive condition is satisfied between point 3 and point 4. The attributes of point 3 and 4 are 1 and 2, respectively, resulting in a difference of 1—thereby fulfilling the difference condition as well.

Now, let us assume that $P_t$ is into $P_t'$ by removing the $i$-th and $(i+1)$-th points and renumbering the points with numbers greater than $i$ by subtracting 2. Then, $P_t'$ becomes identical to $P_{t-1}$. We define such an operation of removal and renumbering as $R_I(P_t, i)$.

Conversely, considering the reversibility from P-data to knot projection, if the following condition is satisfied,

$$\sigma(i) = i + 1,$$
$$R_I(P_t, i) = P_{t-1},$$

it is recognized that a Reidemeister I was applied to the $i$-th segment at time $t-1$. Consequently, the application and location of Reidemeister I can be identified from the transition of P-data.

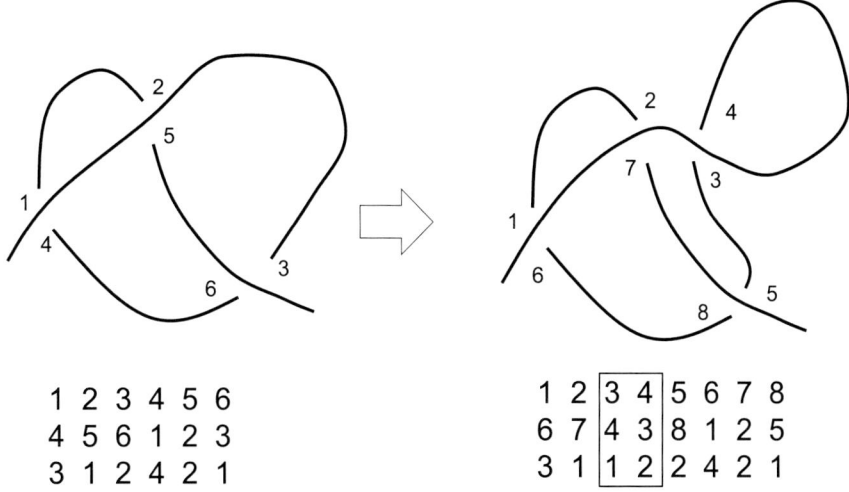

**Fig. 7.4** An example of Reidemeister move I and P-data transition

In Fig. 7.4, removing the third and fourth columns from the right-hand P-data and subtracting 2 from the point numbers of the fifth through eighth points will yield the same P-data as the left-hand P-data. It can be observed that a Reidemeister I move has been applied to the third segment.

### 7.2.3.2 Reidemeister II Moves and Corresponding P-Data Transitions

Figure 7.5 shows an example of a Reidemeister II move. When a Reidemeister II move is applied to the knot projection at time $t - 1$, two intersections, and consequently four points, are added to the knot projection at time $t$. Therefore, the following condition must be satisfied:

$$n(P_t) = n(P_{t-1}) + 4. \tag{7.3}$$

Conversely, if the number of columns of P-data increases by four from $t - 1$ to $t$, it indicates the possible application of a Reidemeister II move.

Assume Reidemeister II is applied to the $i$-th segment and the $j$-th segment, where $i < j$ holds. In the P-data, the two additional intersections will appear consecutively. Therefore, one of the following conditions, either Eq. 7.4 or Eq. 7.5, must be satisfied:

$$\{\sigma(i) = j'\} \cap \{\sigma(i + 1) = j' + 1\}, \tag{7.4}$$

$$\{\sigma(i) = j' + 1\} \cap \{\sigma(i + 1) = j'\}, \tag{7.5}$$

where $j' = j + 2$.

Furthermore, these vertical positions (upper or lower) must be the same, and the signs must be different. Therefore, the following condition must be satisfied:

$$|attr(i) - attr(i + 1)| = 2. \qquad (7.6)$$

Since the state of other intersections remains unchanged, we can remove the $i$-th, $(i + 1)$-th, $j'$-th, and $j' + 1$-th columns from $P_t$ and renumber the point numbers (i.e., subtracting 2 from the point numbers in the range from $i + 2$ to $j' - 1$ and subtracting 4 from the point numbers in the range from $j' + 2$ to $n(P_t)$). This will make $P_t$ identical to $P_{t-1}$. We define such removal and renumbering operation as $R_{II}(P_t, i, j)$.

Conversely, considering the reversibility from P-data to knot projection, if the above conditions are satisfied in P-data, a Reidemeister II move must be applied between the $i$-th and $j$-th segments at time $t - 1$.

Utilizing this fact, the transition of P-data can identify the application of a Reidemeister II move and the segments to which it was applied.

Figure 7.5 illustrates the application of a Reidemeister II move between segment 3 and segment 6. From the P-data,

$$\sigma(3) = 8,$$
$$\sigma(4) = 9,$$
$$|attr(3) - attr(4)| = 2,$$
$$R_{II}P(t|3, 6) = P_{t-1},$$

hold. Therefore, it is confirmed that a Reidemeister II move is applied between the third and sixth segments at time $t - 1$.

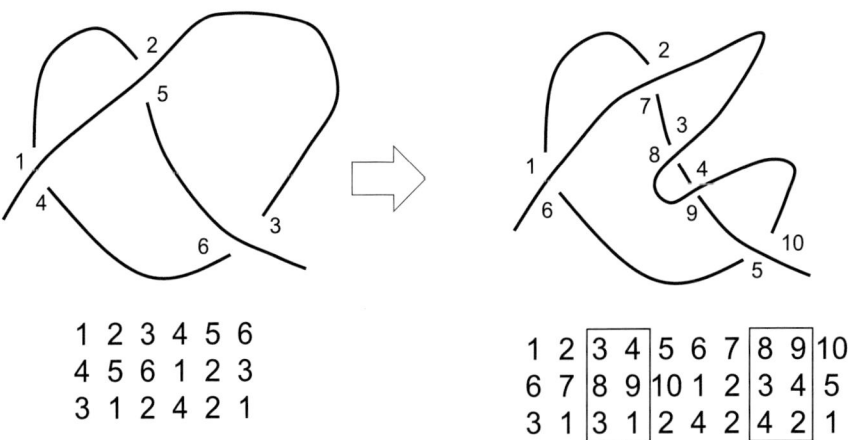

**Fig. 7.5**  An example of a Reidemeister move II and P-data transition

### 7.2.3.3 Reidemeister III Moves and Corresponding P-Data Transitions

When a Reidemeister III move is applied, no intersections are added to the knot projection. That is,

$$n(P_t) = n(P_{t-1}). \tag{7.7}$$

Conversely, if the number of columns of P-data remains the same before and after the deformation action, it indicates the possible application of a Reidemeister III move.

Let us consider the case where the three segments $I$, $J$, $K$, overlap, as shown in Fig. 7.6. In detecting Reidemeister III moves, upper/lower relationships are not considered; therefore, the cases are classified without considering these relationships. The order of $I$, $J$, $K$ can be arbitrary.

There are two ways to assign the numbers to the two points of segment $I$, depending on the enumeration direction on the string. Similarly, there are two ways to assign the numbers to the two points of segments $J$ and $K$. As a result, there are eight possible assignment types as shown in Table 7.1.

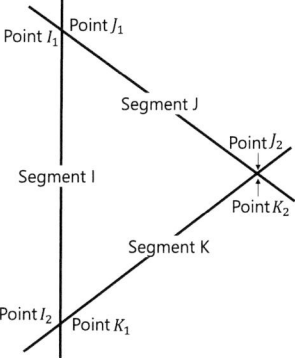

**Fig. 7.6** Definition of segments and points

**Table 7.1** Assignment of point numbers

| | Segment $I$ | | Segment $J$ | | Segment $K$ | |
|---|---|---|---|---|---|---|
| Point | Point $I_1$ | Point $I_2$ | Point $J_1$ | Point $J_2$ | Point $K_1$ | Point $K_2$ |
| Type A | $i-1$ | $i$ | $j-1$ | $j$ | $k-1$ | $k$ |
| Type B | $i-1$ | $i$ | $j-1$ | $j$ | $k$ | $k-1$ |
| Type C | $i-1$ | $i$ | $j$ | $j-1$ | $k-1$ | $k$ |
| Type D | $i-1$ | $i$ | $j$ | $j-1$ | $k$ | $k-1$ |
| Type E | $i$ | $i-1$ | $j-1$ | $j$ | $k-1$ | $k$ |
| Type F | $i$ | $i-1$ | $j-1$ | $j$ | $k$ | $k-1$ |
| Type G | $i$ | $i-1$ | $j$ | $j-1$ | $k-1$ | $k$ |
| Type H | $i$ | $i-1$ | $j$ | $j-1$ | $k$ | $k-1$ |

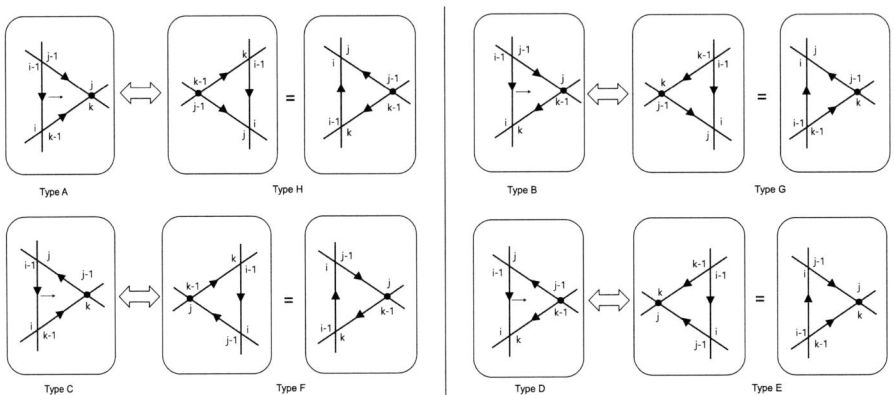

**Fig. 7.7** Transitions in Reidemister move III. The arrows represent enumeration directions along the strings

Considering the possible transitions between these eight types, only eight transitions actually occur between four pairs as shown in Fig. 7.7:

- between Type A and Type H
- between Type B and Type G
- between Type C and Type F
- between Type D and Type E.

Since the state of other intersections remains unchanged, by removing the $(i-1)$-th, $i$-th, $(j-1)$-th, $j$-th, $(k-1)$-th, and $k$-th columns from $P_t$ and $P_{t-1}$, then, $P_t$ and $P_{t-1}$ become identical. We call such an operation $R_{III}(P, I, J, K)$.

Considering the reversibility from P-data to knot projections, the algorithm to determine the procurance and location of a Reidemeister Move III is as follows

- Search for the $I$-th, $J$-th, and $K$-th segments that form a triangle before and after the transition.
- Determine the assignment types for these three segments before and after the transition.
- Verify the validity of the transitions in Fig. 7.7.
- Verify if $R_{III}(P_t, I, J, K) = R_{III}(P_{t-1}, I, J, K)$ holds.

Figure 7.8 is an example of a Reidemeister III move. By focusing on the 3rd, 5th, and 8th segments, the assignment styles at time $t-1$ and $t$ are Type A and Type H, respectively. Indeed, the transition A-to-H is in the list; $R_{III}(P_{t-1}, 3, 8, 5) = R_{III}(P_t, 3, 8, 5)$ holds. Thus, it can be recognized that a Reidemeister III move occurred between the 3rd, 5th, and 8th segments at time $t-1$.

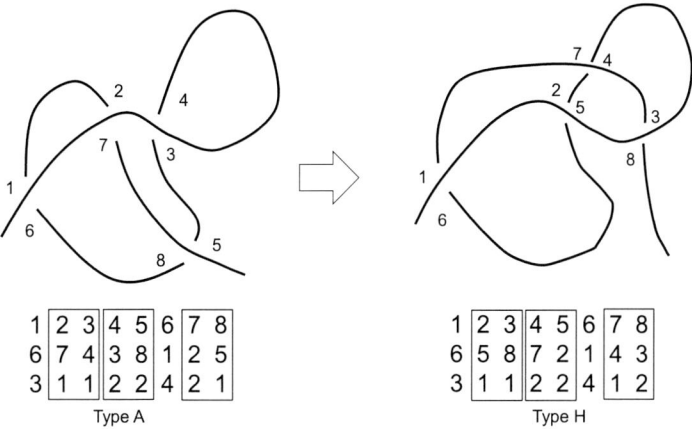

| 1 | 2 | 3 | 4 | 5 | 6 | 7 | 8 |
|---|---|---|---|---|---|---|---|
| 6 | 7 | 4 | 3 | 8 | 1 | 2 | 5 |
| 3 | 1 | 1 | 2 | 2 | 4 | 2 | 1 |

Type A

| 1 | 2 | 3 | 4 | 5 | 6 | 7 | 8 |
|---|---|---|---|---|---|---|---|
| 6 | 5 | 8 | 7 | 2 | 1 | 4 | 3 |
| 3 | 1 | 1 | 2 | 2 | 4 | 1 | 2 |

Type H

**Fig. 7.8** An example of Reidemeister move III and P-data transition

### 7.2.3.4 Cross Moves and Corresponding P-Data Transitions

We introduce a Cross move to deform open curves using one of the end points. Notably, previous analyses of the Reidemeister moves assume that both ends of a string remain stationary to be able to apply the closed-curve theory, Knot theory, to open curves. For the analysis to be practical, it must be capable of generating knots from a single string of an open curve using one of the end points. In this subsection, we introduce a Cross move, which involves translating one end point of an open-curve string across another segment.

Figure 7.3d shows an example of a Cross move. We assume that when a Cross move occurs, no Reidemeister moves occur in other parts, and locally, the end segment creates a new intersection with only one segment of the string. Thus, when this Cross move is applied to the string at time $t-1$, one intersection is added, and the column number of the P-data increases:

$$n(P_t) = n(P_{t-1}) + 2. \tag{7.8}$$

Due to the introduction of this Cross move, when the number of columns in P-data increases by two, two scenarios arise: a Cross move and a Reidemeister I move. The necessity of accounting for these two scenarios represents the inherent cost associated with incorporating the Cross move.

A Cross move is performed by extending the start or end segment across a designated segment. Consequently, the segment to which the Cross move is applied can be determined by identifying the point located either below or above the initial or final point at time $t$, $\sigma(1|P_t)$ or $\sigma(n(P_t)|P_t)$.

When a Cross move is executed between the start segment and the $i$-th segment, the resulting point at the intersection is assigned the number $i+1$. Also, the segment is divided into $I$ and $I+1$ segments.

$$\sigma(1) = i + 1. \tag{7.9}$$

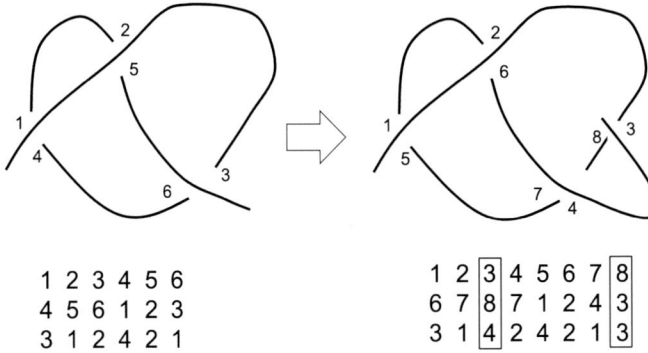

$$
\begin{array}{cccccc}
1 & 2 & 3 & 4 & 5 & 6 \\
4 & 5 & 6 & 1 & 2 & 3 \\
3 & 1 & 2 & 4 & 2 & 1
\end{array}
\qquad
\begin{array}{cccccccc}
1 & 2 & \boxed{3} & 4 & 5 & 6 & 7 & \boxed{8} \\
6 & 7 & \boxed{8} & 7 & 1 & 2 & 4 & \boxed{3} \\
3 & 1 & \boxed{4} & 2 & 4 & 2 & 1 & \boxed{3}
\end{array}
$$

**Fig. 7.9** An example of cross move and P-data transition

In this case, by eliminating the first and $(i + 1)$-th columns from $P_t$ and reassigning the point numbers—specifically, subtracting 1 from the point numbers ranging from 2 to $i$, and subtracting 2 from those ranging from $i + 2$ to $n(P_t)$—$P_t$ and $P_{t-1}$ become identical. This operation is referred to as $C_s(P_t)$.

Next, consider the scenario in which a Cross move is applied between the end terminal and the $j$-th segment. In this situation, the number assigned to the newly introduced point is $j$. As a result, the following condition must hold:

$$
j = \sigma(n(P_t)). \tag{7.10}
$$

In this case, by removing the last and $j$-th columns from $P_t$ and reassigning the point numbers—specifically, subtracting 1 from the point numbers ranging from $j + 1$ to $n(P_t) - 1$—$P_t$ and $P_{t-1}$ become identical. This process is referred to as $C_e(P_t)$.

Conversely, when considering the reversibility from P-data to a knot projection, if $C_s(P_t) = P_{t-1}$, a Cross move must take place between the start terminal and the $(\sigma(1|P_t) - 1)$-th segment. Similarly, if $C_e(P_t) = P_{t-1}$, a Cross move must occur between the end terminal and the $\sigma(n(P_t)|P_t)$-th segment. This framework enables the identification of both the occurrence and location of the Cross move.

In the context of Fig. 7.9, it can be determined that $C_e(P_t) = P_{t-1}$ and $\sigma(n(P_t)|P_t) = 3$. Thus, it is concluded that the Cross move occurred between the end terminal and the 3rd segment.

## 7.2.4  Skill Parameter Determination

Reidemeister and Cross moves require detailed skill information that specifies both where and how they should be applied. The identification of target segments for each move is accomplished through an analysis of P-data transitions. Moreover, specific skill details are necessary for each move, such as determining whether to twist clockwise or counterclockwise, or whether to pass above or below. In polyhedron analysis, these "where-to-apply" parameters are derived by examining the characteristic movements of the object following task recognition through face-contact transitions. Similarly, in string analysis, specific attributes of the P-data are examined to determine these parameters after the identification of each move through P-data transitions.

### 7.2.4.1  Skill Parameters for Reidemeister I Moves

In the case of a Reidemeister I move, there are two scenarios for rotation: twisting either clockwise or counterclockwise. As shown in Fig. 7.10b and c, the clockwise and counterclockwise rotation scenarios are depicted, respectively, relative to the axis lying within the plane and perpendicular to the enumeration direction of Segment $I$. For each twisting direction, a loop forms either on the right or left side, resulting in four distinct cases. In all scenarios, Segment $I$ is divided into three parts, with two new points, $i$ and $i + 1$, introduced at the intersection of the loop.

In the case of a clockwise twist, as shown in Fig. 7.10b, Segment $I$ passes beneath the other part at point $i$. When the loop forms on the right side, the direction of the lower part, relative to the enumeration direction of the upper part, is rightward. Consequently, point $i$ is assigned a minus signature, and the attribute of point $i$ is:

$$att(i|P_t) = 2.$$

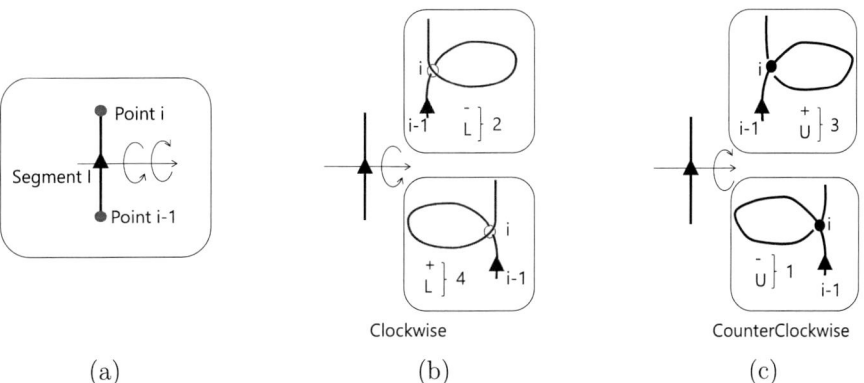

Clockwise                    CounterClockwise

(a)                          (b)                          (c)

**Fig. 7.10** Reidemeister 1 case. **a** Two possible rotations relative to the enumeration direction of segment I. **b** Clockwise rotation. **c** Counterclockwise rotation

When the loop forms on the left side, point $i$ is also located on the lower part. In this scenario, the enumeration direction of the lower part is oriented to the left relative to that of the upper part, resulting in a plus signature. Consequently, the attribute of point $i$ is:

$$att(i|P_t) = 4.$$

Similarly, in the case of counterclockwise twisting, it can be concluded—based on the same reasoning, as shown in Fig. 7.10c—that the attribute of point $i$ is either 3 or 1, depending on whether the loop forms on the right or left side, respectively.

Due to the reversibility of P-data, the reverse process also holds true. Therefore, based on the attribute of point $i$, the side on which the loop forms and the direction of rotation for a Reidemeister I move can be determined.

### 7.2.4.2  Skill Parameters for Reidemeister II Moves

For Reidemeister II moves, loops can form either on the right side (Fig. 7.11a and b) or the left side (Fig. 7.11c and d) of segment $I$, relative to its enumeration direction. Additionally, for each scenario, the enumeration direction of segment $J$ relative to segment $I$ can either align or oppose. These loops result in a total of eight possible configurations, depending on whether segment $I$ passes beneath or crosses over segment $J$.

As shown in the analysis illustrated in the figures, in every case where segment $I$ passes beneath segment $J$, the attribute of point $i$ is an even number, either 2 or 4. Conversely, when segment $I$ crosses over segment $J$, the attribute of point $i$ is an odd number, either 1 or 3. This makes it possible to determine whether segment $I$ passes beneath or over segment $J$ by examining the parity (even or odd) of the attribute of point $i$.

The side (left or right) where the loop forms can be determined from the position of the opposing segment that intersects at point $i$, i.e., $\sigma(i)$.

### 7.2.4.3  Skill Parameters for Reidemister III Moves

Regarding Reidemeister III moves, the vertical relationship among segment $I$, segments $J$, and $K$ is preserved after the move. The transitions in each scenario are uniquely determined, eliminating the need to define a skill parameter.

### 7.2.4.4  Skill Parameters for Cross Moves

Consider a case where segment $I$ applies a Cross move to segment $I$. In this scenario, there are two possibilities regarding the approach direction relative to the enumeration direction of segment $I$. For each of these cases, there are two further possibilities: either segment $I$ passes beneath or crosses over segment $I$. Therefore, as schematically illustrated in Fig. 7.12, a total of four cases will be considered.

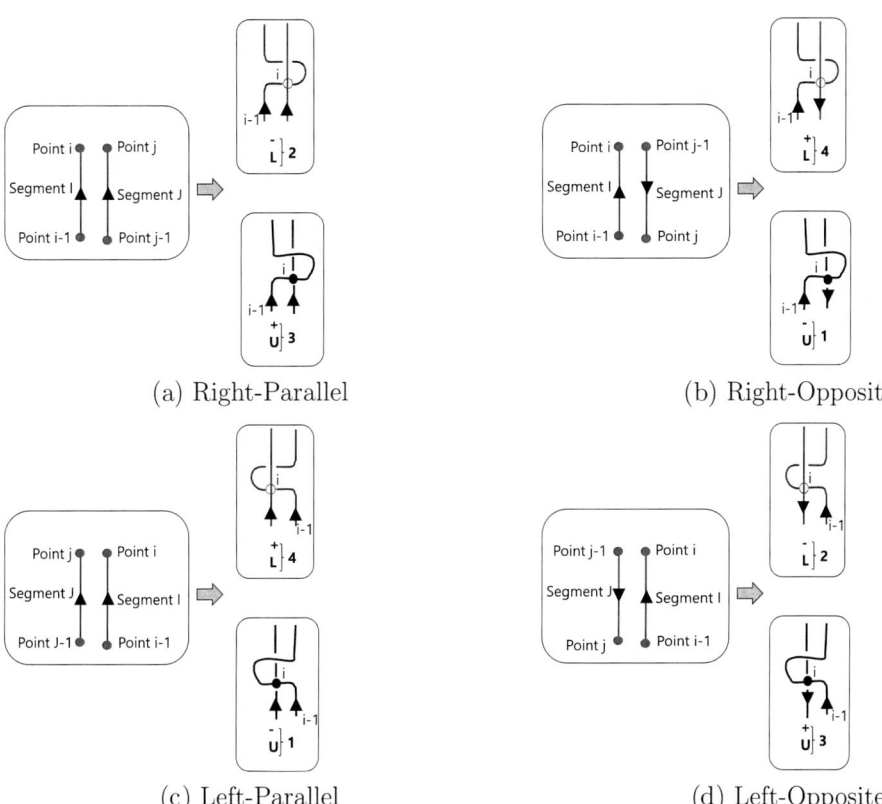

(a) Right-Parallel

(b) Right-Opposite

(c) Left-Parallel

(d) Left-Opposite

**Fig. 7.11** Reidemeister II move

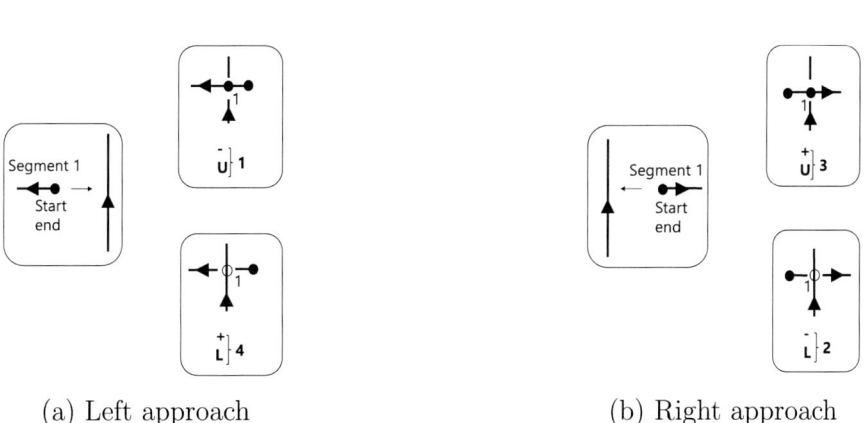

(a) Left approach

(b) Right approach

**Fig. 7.12** Cross move

When crossing over, the attribute values of point 1 depend on whether the approach is from the left or right, relative to the enumeration direction of segment $I$. If the approach is from the left, the attribute value is 1, and if from the right, the attribute value is 3—both odd numbers. When passing underneath, the attribute values of point 1, corresponding to the left and right approaches, are 4 or 2, respectively—both of which are even numbers.

Thus, by examining the parity (odd or even nature) of point 1's attribute values, the vertical relationship can be determined.

## 7.3    Runtime System

### 7.3.1    Task Recognition

From the discussion in the previous section, task recognition—defined as the process of identifying requisite motion skills—can be systematically summarized in Table 7.2. Cases are initially categorized based on the increase in the number of columns in P-data, as demonstrated in the first column. Through the analysis of P-data, the operating segment and the intersecting segment can be derived using the formulas presented in the second column. For instance, in the case of a Cross move depicted in the second row, the operating segment is identified as 1, while the intersecting segment corresponds to $\sigma(i)$. Verification operations are subsequently conducted to validate the determined motions. Finally, the abbreviated names for each identified move are provided in the fourth column of the table.

### 7.3.2    Skill Parameter Determination

Once the required motion skills are determined based on the transition of P-data, skill parameters, such as clockwise or counterclockwise rotations, are determined as follows:

**Reidemeister I move (Rd1)**: Regarding Reidemeister I moves, there are two skill parameters: the direction in which to twist and the direction in which to create the loop. These parameters can be determined based on the attribute value of point $i$, as summarized in Table 7.3.

**Table 7.2** Summary of task recognition

| Increase | Target segment | Verification | Abbreviations |
|----------|----------------|--------------|---------------|
| 0 | $R_{III}(P_t, i, j, k) = R_{III}(P_{t-1}, i, j, k)$ | $R_{III}(P_t, i, j, k)$ | Rd3 (I, J, K) |
| 2 | $\sigma(1)$ | $C_s(P_t)$ | Cr (I) |
| 2 | $\sigma(i)$ | $R_I(P_t, i)$ | Rd1 (I) |
| 4 | $\sigma(i) - 2$ | $R_{II}(P_t, i, j)$ | Rd2 (I, J) |

**Table 7.3**  Skill parameters of Reidemeister I moves

| $attr(i)$ | Loop direction | Rotation |
|---|---|---|
| 1 | Left loop | Counterclockwise |
| 2 | Right loop | Clockwise |
| 3 | Right loop | Counterclockwise |
| 4 | Left loop | Clockwise |

**Reidemeister II move (Rd2)**: For Reidemeister II moves, there is a choice of whether to pass above or below the neighboring segment. This can be determined by examining the parity (even or odd nature) of the point $i$'s attribute, as shown in the table below. The direction in which to create the loop is uniquely determined from the corresponding point of point $i$, $\sigma(i)$. See Table 7.4.

**Reidemeister III move (Rd3)**: For the Reidemeister III move, there is no arbitrariness, and it does not require the determination of parameters.

**Cross move (Cr)**: Regarding a Cross move, there are four cases: whether to pass above or below and from which direction. These are determined based on the attributes of point 1, as summarized in Table 7.5.

**Table 7.4**  Skill parameters of Reidemeister II moves

| $attr(i)$ | Vertical relationship |
|---|---|
| 1 | Over |
| 2 | Under |
| 3 | Over |
| 4 | Under |

**Table 7.5**  Skill parameters of cross moves

| $attr(1)$ | Approach direction | Vertical relationship |
|---|---|---|
| 1 | Left approach | Over |
| 2 | Right approach | Under |
| 3 | Right approach | Over |
| 4 | Left approach | Under |

### 7.3.3 Bowline-knot Analysis

To illustrate the proposed analysis method, we examine the bowline knot as a representative example. The projection of the string is obtained after each motion skill is executed. These projections are subsequently transformed into P-data, as depicted in Fig. 7.13. The proposed method is then employed to identify the requisite motion skill. For clarity and consistency, the process consistently begins the analysis with the same end of the string, marked by a black dot.

**First step**: In the first step, the number of columns in the P-data increases by two, from zero to two. From this, the motion skill that caused this transition has two possibilities: Redemeister I move or Cross move.

- Assuming a Reidemeister I move, the following equations are in fact satisfied:

$$\sigma(1|P_1) = 2,$$
$$R_I(P_1, 1) = P_0.$$

Namely, the cumulative law holds at point 1 and point 2, and the reverse operation of the Reidemeister I move reaches the P-data from the previous step. This confirms that a Reidemeister I move is applied to segment 1.

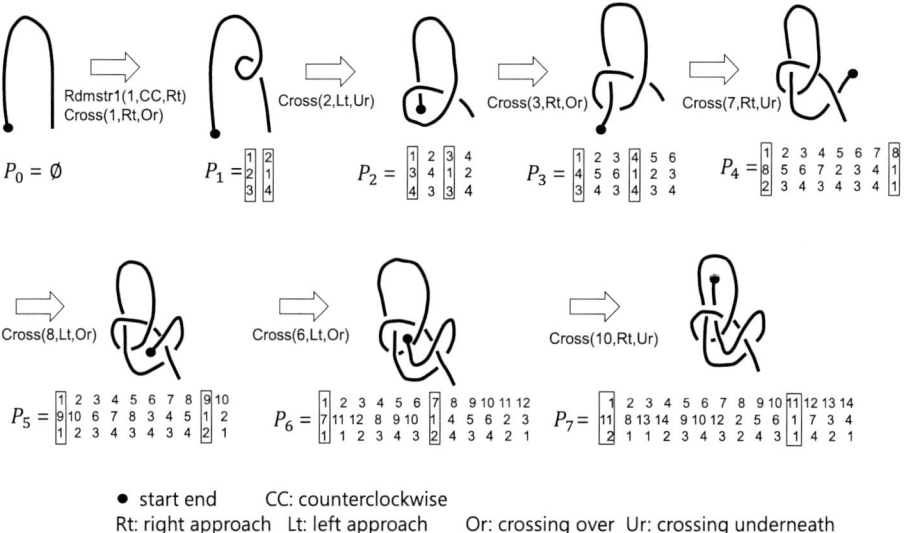

Fig. 7.13 Analysis of bowline knot

Given that a Reidemeister move I is applied to Segment *1*, the attribute of point 1, being 3, provides a skill parameter indicating a counterclockwise rotation (CC) on the right side (Rt) in the enumeration direction of segment 1 from Table 7.3.

- Assuming a Cross move, the following equation also holds:

$$\sigma(1|P_1) = 2,$$
$$C_s(P_1) = P_0.$$

The reverse operation of the Cross move also reaches the P-data from the previous step. This also confirms that a Cross move is also applied to segment 1.

Given that a Cross move is applied to Segment 1, the attribute of point 1, being 3, provides a skill parameter that indicates a cross-over move (Or) of the starting point from the right direction (Rt) with respect to the enumeration direction of segment 1 from Table 7.5.

This ambiguity arises due to the introduction of the Cross move to apply Knot theory, initially developed for closed curves, to the context of typing of an open-curve string. Nevertheless, fortunately, the same result is achieved, regardless of which motion skill is employed.

**Second step**: In the second step, as the column numbers of P-data increase from two to four, there are, once again, two possible motion skills that could be applied:

- Assuming a Reidemeister I move, commutative law should hold at two consecutive points. However, no such pair is found in $P_2$. Therefore, this assumption is rejected.
- Assuming a Cross move, the following equation holds,

$$\sigma(1|P_2) = 3,$$
$$C_s(P_2) = P_1.$$

Hence, it can be concluded that the Cross move can be applied to segment 2.

Given that a Cross move is applied to Segment 2, the attribute of point 1, being 4, provides a skill parameter that indicates a crossing-underneath (Ur) from the left side (Lt) with respect to the enumeration direction of segment 2.

**The remaining steps**: By conducting similar analyses, it becomes possible to identify the specific motion skills and corresponding skill parameters that should be applied to each segment, as illustrated in Fig. 7.13. In this particular example, the majority of transitions resulted in Cross moves. This could be attributed to the fact that, when transforming open curves, manipulating the endpoints tends to be a comparatively simpler operation.

## 7.3.4  System

The system, as is typical in LfO, is divided into a demonstration phase and an execution phase.

### 7.3.4.1 Demonstration Phase

During the demonstration phase, the demonstrator utilizes the stop-and-go method to perform string manipulation. At each stage of the knot-tying process, the demonstrator deliberately pauses to allow the system to observe and record the current state of the string. This step-by-step procedure is repeated systematically until the final knot is completed.

For each stage of the demonstration, the following process is repeated, gradually generating the required sequence of motion skills.

- **Image processing**

  - **Image acquisition** acquires RGB and depth images of the string.
  - **Background subtraction** subtracts background data from the images to extract only the string pixels.
  - **Thinning** performs thinning of the string pixels using the Hilditch filter and removes errors.
  - **Graph extraction** identifies intersections and endpoints of the string by counting adjacent points for all pixels in the thinned image. Additionally, it enumerates all segments.

- **P-data generation**

  - **Enumeration** performs the numbering of the intersection points and segments from the start point to the end point of the knot.
  - **Orientation** determines whether the vertical positions of the intersection points are upper or lower.
  - **Compilation** derives P-data from the intersection point numbers, enumeration directions, and vertical positions.

- **Task recognition**
  This process identifies the corresponding motion skills based on the transitions of the obtained P-data. In cases where multiple motion skills are applicable, the selection is made according to a predefined priority order, with a Cross move taking precedence, followed by a Reidemeister move.

- **Parameter determination**
  This process determines the corresponding skill parameters based on the identified motion skill. The analysis primarily relies on the attributes of a specific intersection point.
  At this stage, running parameters, such as where to grasp, can also be obtained from the thinning images. For further details, please refer to [11].

### 7.3.4.2 Execution Phase
In the execution phase, the robot converts each motion skill into robot commands to create the knot. Figure 7.14 shows an example of the execution.

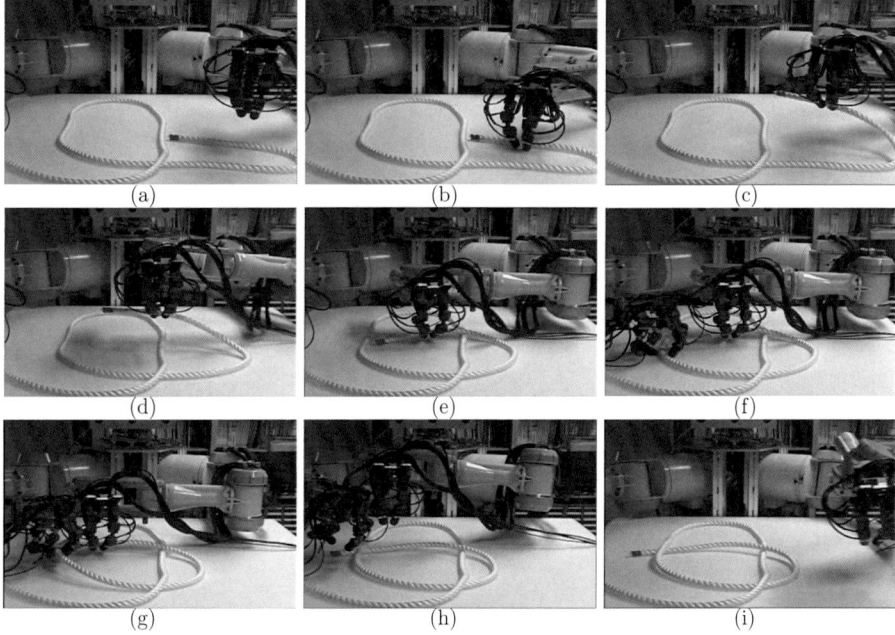

**Fig. 7.14** Robot execution of knot-typing (a cross move skill)

# Dance World

**8**

This chapter delves into robot dance as a prime example of directly mimicking human movements. In prior chapters, the objective of Learning from Observation (LfO) was not to replicate human movements, but to generate robot movements that achieve the same outcomes, such as identical face contacts, as those produced by the human movements. This approach addresses the hardware disparities between humans and robots. In contrast, this chapter introduces an LfO framework aimed at generating robot movements that are perceived by the audience as similar—though not necessarily identical—while addressing hardware discrepancies. This approach leverages Labanotation, a representation originating from the dance community.

The Labanotation employed in the dance community is used to describe the essence of movements. Dance involves the entire body, making direct mimicry extremely difficult due to various factors such as the functional differences between robots and humans, as well as differences in weight and height. Thus, to replicate these movements, it is necessary to capture only the essence of the movements within the permissible range of the specific robot hardware. Interestingly, the same situation arises between a human dance teacher and a student. Teachers and students have different heights and weights, making it impossible for students to replicate all of the teacher's movements. Therefore, students must imitate only the fundamental parts of the dance—the parts that appear the same to the audience—and adjust the other parts according to their own physique. To this end, the dance community devised a method called Labanotation to describe the essential parts of the dance, and dancers are trained to express only these key points with their bodies.

The system introduced in this chapter represents robot tasks centered around Labanotation. Human movements are first encoded into Labanotation. Subsequently, each Labanotation is decoded and executed by the robot. Different methods are employed for mimicking the

K. Ikeuchi et al., *Learning-from-Observation 2.0*, Synthesis Lectures on Computer Vision,
https://doi.org/10.1007/978-3-032-03445-8_8

upper body and the lower body, with the waist serving as the central point. The movements of the upper body, particularly the arms and hands, are distinctive features of dance. Therefore, these upper body movements are implemented as individual tasks using the Labanotation employed in the dance community. Robot movements are generated through skills that map each Labanotation to corresponding upper body movements.

Regarding the lower body, in some dance forms, leg movements are crucial. However, in the case of the traditional Japanese dance studied here, the legs are often concealed by clothing. Experimental data from motion capture reveal that stride length and steps vary with the tempo of the music. Analysis shows that stability and movement distance are prioritized over expression. Therefore, for the lower body, the focus is on the robot's stability. Attention is given only to support information for landing points in Labanotation, while gesture information in Labanotation is ignored.

## 8.1    Upper Body Representation

### 8.1.1   Labanotation

This section first introduces Labanotation and then describes the method for expressing the movements of the upper body in humanoid.

#### 8.1.1.1 Definition of Labanotation

Labanotation is a notation method of recording dance movement developed by Rudolf Laban in the early 20th century [12]. It consists of three elements: the notation of body parts, motion directions of the body parts, and the durations of the motions. The Labanotation score is drawn in two dimensions, with the horizontal axis representing the body parts and the vertical axis representing time, as shown in Fig. 8.1a, where time flows from bottom to top. Symbols, such as rectangles or triangles shown in Fig. 8.1a, are arranged in columns corresponding to each body part. The shape and texture of each symbol indicates the direction of posture for the corresponding body part at that time. The length of a symbol is adjusted to fit the start and end times of each movement. The gap between two symbols for the same body part indicates that there is no movement during that period and the previous pose is maintained.

Columns are divided into left and right, corresponding to the left and right sides of the body. These vertical columns, which represent body parts, are indicated by arrow-like symbols, as shown in Fig. 8.1b. This example illustrates that these columns correspond to the upper and lower arms on both sides of the body, as well as the head. The four central columns represent the legs that support the body. Two of these columns are designated as support columns, indicating the legs that bear weight, while the remaining two are known as gesture columns, which indicate movements made with the legs without bearing weight. Labanotation does not necessarily need to specify all body parts. For instance, when the focus is on upper body movements, as in this sub-section, only the movements of the upper

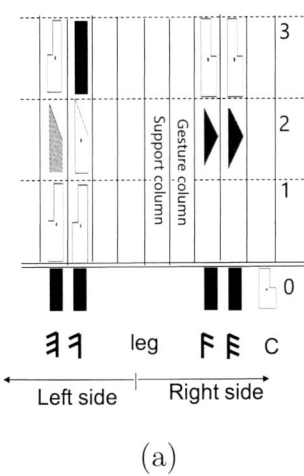
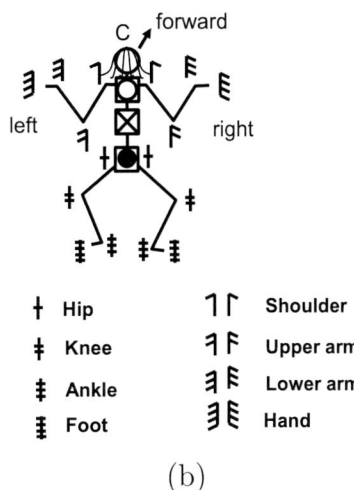

Fig. 8.1 Labanotation

arms, forearms, and head on both sides are explicitly represented, while other body parts, such as support and gesture information for the left and right legs, can be omitted.

Labanotation represents the movement of body parts relative to the overall coordinate system of the body. In Labanotation, the main coordinate system, shown as the XYZ axes in Fig. 8.2a, is first defined. The origin of the main body coordinate system (known as the "main cross" in the dance community) is at the center of the body when the dancer stands naturally. The X-axis (front-back) indicates the direction the dancer is facing, the Y-axis (left-right) indicates the direction from right to left, and the Z-axis (up-down) indicates the direction from the feet to the head. Each body part has a local coordinate system parallel to the body coordinate system, defined at the joint closer to the center of the body part. The X' Y' Z' axes in Fig. 8.2a are an example of a local coordinate system defined for the left upper arm, with the origin at the left shoulder (the joint closer to the center of the left upper arm).

In Labanotation, symbols such as trapezoids and triangles represent the horizontal directions of movement for each body part. Based on the local coordinate system, these symbols indicate eight directions: front (X-axis), back (-X-axis), left (Y-axis), right (-Y-axis), front-left, front-right, back-left, and back-right. For forward and backward directions, two symbols are used based on the specific body parts pointing forward and backward. For example, the left elbow uses the left symbol for forward direction, while the right elbow uses the right symbol for forward direction. Additionally, rectangular symbols indicate the zenith and nadir in the Z-axis direction, referred to as "place." Consequently, eleven symbol shapes are utilized as shown in Fig. 8.2b.

The shading of each symbol indicates the zenith angle direction. Figure 8.2c illustrates the shading corresponding to the zenith angle direction, indicating high, medium, and low.

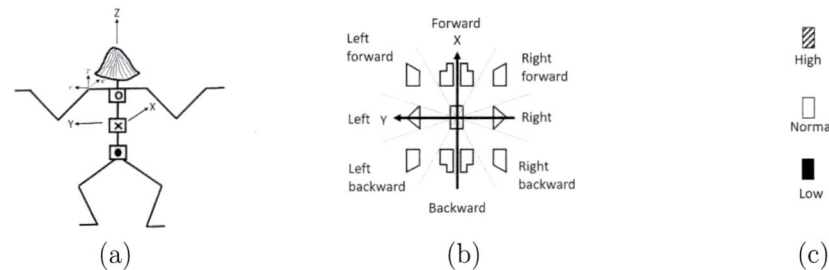

**Fig. 8.2** Coordinates in Labanotation

For the zenith and nadir, shading is added to the place symbols (rectangular shapes) to indicate high and low. Thus, the zenith angle direction includes five directions: zenith, high, medium, low, and nadir.

The length of the symbols represents the elapsed time from the start to the end of each movement. The bottom edge of the symbol indicates the start time, while the top edge indicates the end time.

In summary, each symbol in the score contains two types of information: the final pose of each body part and the elapsed time to reach it. It is important to note that Labanotation symbols do not describe the trajectory between the starting and ending poses.

As an example, let's interpret the Labanotation score in Fig. 8.3e. Initially, in the preparatory state below the double lines, both arms are pointing down, indicated as place and low as shown in Fig. 8.3a. In the first segment, the left upper and lower arms are raised forward, while the right arm remains pointing down, as shown in Fig. 8.3b. In Labanotation conventions, the representation of the right side, which did not move, is expressed as a blank in the score. In the second segment, the left upper arm points horizontally to the left forward, and the left lower arm points high to the left forward, Meanwhile, the right upper and lower arms

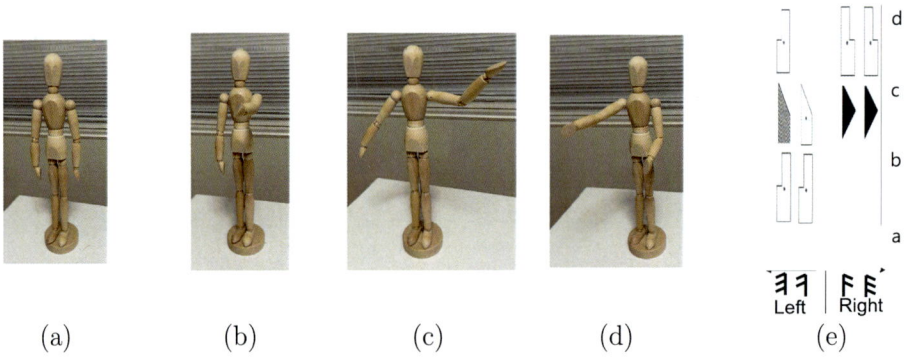

**Fig. 8.3** Upper-body poses and corresponding Labanotation

point down to the right. See Fig. 8.3c. In the third segment, illustrated in Fig. 8.3d, the left upper arm points down, the left lower arm points forward, and the right upper and lower arms point forward at a medium height. As shown in this example interpretation, Labanotation can be utilized to indicate the pose of the upper body at each moment.

In our system, achieving a pose represented by a single Labanotation symbol is defined as a task. Each Labanotation symbol corresponds to a state (pose) transition, where the ending state (pose) is depicted by the Labanotation symbol. The starting state is the pose depicted by the previous Labanotation symbol. The time required for the task, or the skill parameter, is represented by the length of the symbol. Therefore, if a series of Labanotation symbols can be extracted from a sequence of human movements, task recognition for the upper body is considered complete.

### 8.1.1.2 Justification of Labanotation

Labanotation scores are conceptually analogous to musical scores. Just as skilled musicians can transcribe the music score by listening to a musical performance, dance experts can record movements as Labanotation scores by observing a dance performance. Reading the musical score allows musicians to perform a piece so that the audience recognizes it as the same music, albeit with slight variations each time. Similarly, interpreting a given Labanotation score enables dancers to perform a dance such that the audience perceives it as the same dance, despite minor variations in each performance.

Using Labanotation, one can determine whether dance performances belong to the same set of equivalent dances. The necessary and sufficient condition for two dance performances to belong to the same set is that they can be described using the same Labanotation score.

- **Sufficient condition**: According to the Labanotation Committee, if two dance performances belong to the same set of equivalent dances, they are recorded using the same Labanotation score.
- **Necessary condition**: When observing the same Labanotation score, various dances can yield different performances. However, the committee ensures that these performances are recognized as the same dance by human dancers.

In practice, to fine-tune the concept of equivalent dance sets, the Labanotation community has established training courses for recording dances. They issue certificates as dance recorders only to those who pass the exam.

The eight divisions for azimuth and five divisions for zenith angles in Labanotation may appear somewhat coarse, but psychological evidence supports their adequacy. Miller suggested that humans can process analog information into approximately plus or minus two categories [24]. For instance, our eyes perceive continuous color changes in a rainbow as seven distinct colors, and there are precisely seven main chords in music. When evaluating performances, a five-point scale—excellent, good, fair, poor, and bad—is often employed. Based on Miller's assertion, the digitization into eight directions for azimuth and five directions for zenith angles is reasonable, considering the limits of human perception. The fact that the dance community has utilized this notation system for over a century further substantiates this point. Moreover, if necessary, Labanotation allows for finer directional specifications when necessary by redefining specific directional ranges and digitizing them into approximately seven directions within those ranges.

## 8.1.2  Labanotation Extraction

This section outlines the method for extracting Labanotation from human movements, which involves two steps: keyframe detection and encoding. Firstly, it is essential to identify when significant poses occur within the sequence of human movements. These moments, when poses are recorded, are referred to as keyframes. Subsequently, the poses at these keyframes are encoded into Labanotation symbols.

### 8.1.2.1 Keyframe Detection

The sub-section explores the method for determining the timing of the encoding, referred to as keyframes. For keyframe extraction, we have developed three methods[1] the Naive method, the Total Energy method, and the Parallel Energy method [13]. Each method has its advantages and disadvantages, so we will introduce all three here. The Naive method is simple and easy to understand. The Total Energy method is robust to noise, performs consistently, and is most suitable for general motion analysis. Conversely, for subtle movements such as those in dance, detailed processing using the Parallel Energy method is required. However, it may generate additional frames, making it desirable to be used in combination with music.

**Naive method**: This method encodes human poses into Labanotation at each sampling point without determining whether it is a keyframe or not. By comparing the current Labanotation symbol with the one from the previous sampling, any difference indicates that this sampling point is a keyframe.

  As a result of implementing this method, the number of detected keyframes was several times higher than the number recorded by Labanotation experts. Additionally, it was found that generating a human-like Labanotation score using this method is challenging. The issues can be summarized into the following three points:

- When a single movement of a body part spans multiple Labanotation regions, multiple Labanotation symbols may be generated to represent that movement. For example, if a body part moves from a 'place low' (the South Pole) to a 'place high' (the North Pole),

---

[1] These methods are available as open-source from Microsoft. https://github.com/microsoft/LabanotationSuite.

the system generates multiple symbols to map the trajectory between these two points. However, for robotic execution, the focus is solely on the stopping points at the start and end of the movement, without requiring deceleration at intermediate locations, such as the equator.

- Even during the generation of a single pose, different body parts may cease movement at slightly varying times. This variability, inherent in human performance, leads to the creation of multiple Labanotation symbols corresponding to those timings. As a result, a single meaningful movement may produce multiple keyframes.
- Labanotation converts continuous space into digital representations by applying angular thresholds. However, during this process, noise near the detection thresholds can lead to the generation of extra Labanotation symbols. As a result, these additional symbols are produced close to the detection threshold.

This naive method retains intermediate trajectory information as a symbolic representation, providing an advantage despite being less capable of generating meaningful Labanotation compared to human experts. Such symbolic representation can be beneficial for skill analysis and serves as auxiliary information for the Parallel Energy method discussed next.

**Total energy method**: We developed a method that analyzes the speed of body parts, incorporating insights from discussions with Labanotation experts. These experts highlighted the importance of short pauses in body movements, which significantly influence the perception and expression of human dance. Dancers consciously integrate these pauses into their performances, emphasizing their relevance. In Japanese Noh performances, short pauses, referred to as 'tome,' are particularly crucial for expressive elements. Furthermore, experts suggest that humans perceive continuous movements as sequences of poses defined by these pauses. Based on this understanding, our total Energy method is designed to extract these brief pauses as keyframes.

The total energy method is solely based on the speed of hand movement, independent of music. Although keyframe representation in dance is typically synchronized with the beat of the music, experts are capable of extracting keyframes without relying on musical cues. Adopting a music-independent approach broadens the scope of applications beyond dance analysis. As a result, we calculate the energy function exclusively using the speed and acceleration of the hand's position, $(x(t), y(t), z(t))$, and identify keyframes by detecting the local minima of this function. Notably, these hand positions are represented with respect to the body coordinate system.

Joint angle positional data captured by a sensor such as Kinect or motion capture systems often includes noise. To mitigate this, a Gaussian filter is applied to the trajectories of the left and right hand positions, effectively smoothing the input data.

The total energy method utilizes the combined energy of the left and right hands. The total energy function for the hand is derived using both velocity and acceleration:

$$E(t) = g(E_a(t)) - k(E_s(t)), \tag{8.1}$$

where $E_a$ represents the acceleration calculated as

$$E_a(t) = \frac{1}{\sqrt{3}} \sqrt{\left(\frac{\partial^2 x(t)}{\partial t^2}\right)^2 + \left(\frac{\partial^2 y(t)}{\partial t^2}\right)^2 + \left(\frac{\partial^2 z(t)}{\partial t^2}\right)^2}. \tag{8.2}$$

$E_s$ is the speed determined as

$$E_s(t) = \frac{1}{\sqrt{3}} \sqrt{\left(\frac{\partial x(t)}{\partial t}\right)^2 + \left(\frac{\partial y(t)}{\partial t}\right)^2 + \left(\frac{\partial z(t)}{\partial t}\right)^2}. \tag{8.3}$$

Additionally, $g$ and $k$ are normalization functions described below.

The energy term $E_a$ and $E_s$ are normalized to a range of $[0, 1]$ based on the maximum and minimum values observed throughout the entire observation period.

$$g(E_a(t)) = \frac{E_a(t) - E_{a-min}}{E_{a-max} - E_{a-min}}. \tag{8.4}$$

$$k(E_s(t)) = \frac{E_s(t) - E_{s-min}}{E_{s-max} - E_{s-min}}. \tag{8.5}$$

These values are calculated for both the left and right hands, and their sum provides the total energy.

$$E_{total}(t) = \sum_{left,right} g(E_a(t)) - \sum_{left,right} k(E_s(t)). \tag{8.6}$$

The total energy method detects peaks in the total energy function, $E(t)$, and designates those timings as keyframes.

**Parallel energy method:** In certain types of human movements, particularly in dance, each body part moves independently and in parallel. As a result, the lengths of the Labanotation symbols in each column can vary from those in other columns. For instance, as shown in Fig. 8.2a, while the left lower and upper arms are raised to an intermediate height, the right lower and upper arms remain in the same position as in the initial keyframe. This demonstrates that each body part is associated with its own keyframe. Therefore, to facilitate detailed dance analysis, it is essential to prepare a distinct energy function for each body part.

The Parallel Method evaluates motion using velocity and acceleration in polar coordinates. In contrast, the previous total energy method relied on Cartesian coordinates, as it focused on the speed and position of the hand, which represents the terminal part of the body. However, the parallel energy method, which evaluates the movement of individual body parts, adopts a more appropriate approach by using the local spherical coordinate system defined at each body part. This is because each body part primarily moves through rotational motion around the joint axes, maintaining a fixed distance between joints.

The position data of each joint is initially expressed in the Cartesian world coordinate system and must be converted to a local polar coordinate system. To achieve this, the joint's position in the Cartesian world coordinate system is first transformed into a local Cartesian coordinate system as $(x(t), y(t), z(t))$. The origin of this local Cartesian coordinate system, which also serves as the origin of the corresponding local polar coordinate system, is positioned at the joint closer to the center of the body. Subsequently, the coordinates in the local Cartesian system are further represented in the local polar coordinate system as follows:

$$r(t) = \sqrt{x(t)^2 + y(t)^2 + z(t)^2}, \tag{8.7}$$

$$\theta(t) = \cos^{-1}\left(\frac{z(t)}{r(t)}\right) \quad \theta(t) \in [0, 180], \tag{8.8}$$

$$\phi(t) = \tan^{-1}\left(\frac{y(t)}{z(t)}\right) \quad \phi(t) \in [-180, 180]. \tag{8.9}$$

We calculate the geodesic angular speed for all four upper-body parts (right upper and lower arms, left upper and lower arms) as follows:

$$\mathbf{v}(t) = \left(\frac{\partial \phi(t)}{\partial t} \sin \theta(t)\right) \mathbf{e}_\theta(t) - \frac{\partial \theta(t)}{\partial t} \mathbf{e}_\theta(t). \tag{8.10}$$

We define the energy function of each body part as its speed:

$$E_p(t) = |\mathbf{v}(t)|. \tag{8.11}$$

We employ two Gaussian filters with different $\sigma$ to smooth the result of the energy function and to detect the precise positions of key frames.

$$E_{partial}(t) = E_p(t) * G(x(t)). \tag{8.12}$$

To begin, the presence of global minima is identified using the large $\sigma$ filtered data. Next, scale-space theory [181] is applied to refine the approximate locations of these minima using the small $\sigma$ data. Specifically, the search range is defined as the interval between the two inflection points that surround the global minima in the large $\sigma$ data. Within this range, the precise location of the minima is determined by identifying the minimum value in the small $\sigma$ data.

### 8.1.2.2 Labanotation Encoding

In the Parallel Method, keyframes are detected for each body part's movement, which makes their grouping necessary. Due to slight variations in the movements of each body part, the timings that should ideally correspond to the same keyframe may differ subtly across parts, resulting in multiple keyframes around a specific timing. Therefore, grouping these keyframes is essential to maintain consistency.

The Parallel Method requires the identification of intervals of no movement, referred to as holding intervals. In contrast, the previous Total Energy Method defined keyframes based on the total energy, which was determined by the overall movement of the body. For non-moving parts, keyframes were derived from the total energy generated by other moving parts, allowing these keyframes to identify the intervals during which the non-moving parts remained in a holding position. However, in the Parallel Method, holding intervals must be identified for each body part individually, using its specific energy function. The boundaries of these intervals are then defined as keyframes.

The process begins by identifying the local minima across the entire interval. Next, the movements of the energy function within the intervals defined by these minima are analyzed to determine whether they represent holding intervals or movement intervals. Finally, the local minima of body parts in close proximity are grouped together, and the keyframes to be converted into Labanotation are finalized.

**Holding detection**: Holding refers to intervals during which a body part remains in a specific pose, depicted as a blank space in Labanotation. Over the course of a holding interval, the energy function typically exhibits a characteristic pattern: it starts with a decrease in energy caused by a reduction in the part's speed, shifts into a low-energy phase that signifies a period of rest, and ends with an increase in energy as the part transitions back into motion.

To detect intermediate intervals, a naive Labanotation generated by a simple method is utilized, effectively eliminating peaks caused by subtle dancer movements. Human dancers often find it challenging to keep their joints completely still. As a result, during holding intervals, energy values fluctuate around zero due to minute movements, leading to small peaks and valleys. While it is possible to set a threshold to filter out these peaks, the arbitrary nature of thresholds is undesirable. Therefore, a continuous Labanotation generated by the naïve method is employed to avoid this arbitrariness.

The determination of intervals relies on whether the same Labanotation is consistently maintained. During these intervals, dancers preserve the direction specified by a particular Labanotation. This direction is assumed to lie within the central region of the spatial directions defined by Labanotation. Thus, even if a dancer's movements exhibit subtle fluctuations, these remain within the central region of the specified Labanotation range and do not cross into other spatial direction intervals, maintaining continuous Labanotation without changes.

An interval is classified as holding when successive frames exhibit the same naïve Labanotation with multiple valleys. In such cases, only the first and last valleys are retained as the starting and ending frames, while any intermediate valleys are disregarded.

**Brief Stop detection**: After processing the holding intervals, the remaining local minima are analyzed to identify the keyframes. These identified local minima are then designated as the keyframes.

**Auxiliary frame insertion**: The Labanotation committee specifies that when the angle between two keyframes exceeds $135°$ and the trajectory deviates from a geodesic path, an auxiliary frame must be introduced at the midpoint of the trajectory. Consequently, in such instances, an auxiliary frame is added at the midpoint.

**Grouping**: In the Parallel Method, after obtaining partial keyframes for each body part, their positions are adjusted and grouped. Due to human physical limitations and potential measurement inaccuracies, achieving an overall pose, represented as a single set of Labanotation, may lead to slight variations in the timing at which different body parts reach their final positions or are recorded as such. As a result, these partial keyframes are grouped and collectively referred to as body keyframes.

Dancing is often performed in sync with music; hence, while the music beat itself is not directly used, a 1/3 s window—approximately 25% of the smallest beat in music—is adopted for grouping. This window is scanned along the timeline, clustering partial keyframes within the window into a single body keyframe. The timing of partial keyframes to be grouped into the same body keyframe is assumed to be distributed around the timing of the original body keyframe. Additionally, it is considered that the timing of each body keyframe is relatively distant from those of other body keyframes. Finally, the average position of the partial keyframes within each group is calculated and used as the timing for the body keyframe.

**Spatio-encoding**: In the Parallel Method, the same spatial partitioning as the Total Method is employed. The eight directional orientations and five zenith directions in Labanotation are represented as 26 distinct directions on a Gaussian sphere, as illustrated in Fig. 8.4. The conversion of poses in each keyframe into Labanotation is determined by the angular directions of individual body parts. Based on the measured geodesic distance, the symbol corresponding to the nearest sampling direction among the 26 directions is assigned for the respective Labanotation.

### 8.1.3  Upper-Body Task Model

#### 8.1.3.1 State and Task

Dance tasks, based on the aforementioned discussion of equivalence, can be defined as creating poses represented by Labanotation at specific moments in time, referred to as keyframes. Within the LfO framework, the state is defined as poses expressed through Labanotation or those within its representational scope. The task, on the other hand, refers to the transitions leading to the final poses represented by Labanotation or the corresponding movements of the robot required to achieve those poses.

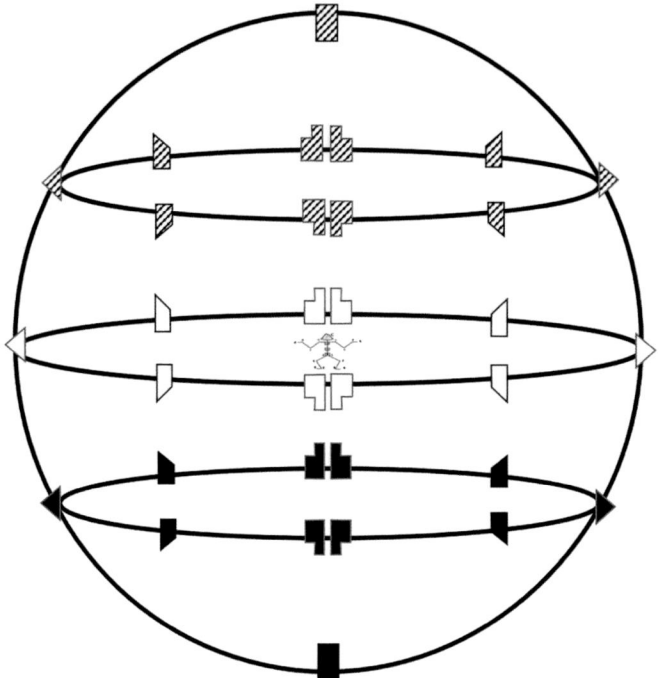

**Fig. 8.4** Sampling directions in Labanotation

Unlike tasks in the polyhedron LfO or the knot-tying LfO, where the initial state is included as part of the task definition, dance tasks are defined exclusively based on the state following the transition. This distinction is particularly evident in upper-body movements in dance, where the absence of environmental constraints on the movements themselves makes it unnecessary to account for restrictions at the initial state. Instead, attention is focused solely on the final state. This approach offers advantages, such as reducing the number of tasks requiring preparation. Moreover, once Labanotation has been successfully extracted from human movements, it can serve as a complete means for recognizing upper-body tasks, providing further efficiency.

### 8.1.3.2 Labanotation Mapping

To replicate upper-body movements, it is necessary to calculate the robot's poses corresponding to Labanotation at each specific moment. This calculation can be performed relatively easily using Forward Kinematics (FK). Directions for the upper arm and forearm are mapped in advance using FK, with shoulder and elbow joints as the reference points. If the robot has a higher degree of freedom (DOF) compared to the Labanotation score, linked directions are mapped to two parts of the robot. Conversely, if the DOF of the Labanotation score exceeds that of the robot, adjacent Labanotation symbols are recursively combined into a single symbol until the DOF matches that of the robot. The depth of the recursive operation is set to three [13].

Robot movements require a trajectory to specify intermediate motions. The initial pose is derived from the final pose of the previous task, while the final pose is obtained through mapping from Labanotation. Joint angles between these poses are determined through linear interpolation. This result is then passed to the full-body motion generation module, which will be discussed later. Although hierarchical representation methods utilizing intermediate points [182] were considered to enhance the aesthetic effect of dance, their details are omitted here.

## 8.2   **Lower Body Representation**

Replicating human lower body movements directly with a robot presents significant challenges. While motion capture technology can provide accurate data on landing points and trajectories, humanoid robots—though designed to resemble the human body—differ substantially in joint structures (degrees of freedom and range of motion), body shapes, compositions, and physical capabilities (such as joint actuator performance). These disparities make it difficult for robots to execute movement trajectories identical to those of humans.

This challenge becomes even more pronounced in dance performances involving bipedal robots, as they must not only replicate the upper body's movements but also support their own weight. Human feet are capable of flexible, shock-absorbing landings from heel to toe, enabling stable motion. Conversely, current bipedal robots often feature rigid, plate-like feet with limited degrees of freedom, which restrict their ability to achieve stable landings and provide adequate support. These limitations further complicate the task of accurately reproducing lower body movements in robotic systems.

To address this challenge, we developed lower body task models based on Labanotation. Each task model is designed with skills that enable the execution of lower body movements while considering the physical characteristics of the robot, such as its trajectory. During the demonstration phase, task sequences are derived from motion capture data, and corresponding skill sequences are executed accordingly. To prevent falls caused by upper body movements, various real-time filters are also introduced. By employing this framework, we aim to effectively address the strict constraints placed on leg movements, ensuring that the robot can stably reproduce lower body movements while preserving the characteristics of the original motions.

In the following sections, we will first explain the lower task models and describe how they are used to generate robotic movements. Subsequently, we will discuss the specific challenges associated with reproducing dance movements in bipedal robots and outline the on-line filtering methods for generating stable movements.

### 8.2.1   Lower-Body Task Model

#### 8.2.1.1 Support Column and Gesture Column

The movements of the lower body can also be represented using Labanotation. Specifically, for each leg, two columns are prepared: the support column and the gesture column. The support column mainly concerns the timing and position of the legs' landing, while the gesture column deals with the gesture of the free leg in the air.

In the reproduction of Japanese dance, we have decided to focus solely on the support column in Labanotation. Unlike Western dance, where the lower body plays a crucial role and is required to follow Labanotation with an emphasis on lower body gestures, Japanese dance traditionally limits the visibility of lower body movements due to the practice of performing in kimonos. As a result, the lower body gestures are less significant for the visual expression of Japanese dance.

Furthermore, research has shown that while the stride length and trajectory of the lower body, represented in the gesture columns, adjust with the tempo, the number of steps, represented in the support columns, remains consistent, regardless of tempo [183]. These observed changes and consistencies seem to prioritize the stability of the lower body over its expressiveness. Consequently, in this reproduction of Japanese dance, the gesture column is disregarded, and only the support column is utilized in Labanotation.

#### 8.2.1.2 State and Task

In terms of lower body movements, the task is also defined as one that causes a transition in states. Similar to the approach taken for the upper body, the state of the lower body is determined based on the condition of the feet, described in the support column of Labanotation.

The states in the support columns are categorized into three possible cases: both columns contain symbols, only one column contains a symbol, or neither column contains a symbol (refer to Fig. 8.5a). The STAND state, where symbols are present in both columns, corresponds to a pose supporting the body with both feet grounded. In contrast, the STEP state occurs when a symbol is present in either the right or left column, indicating a pose where one foot is lifted and then re-grounded. This is further classified as L-STEP and R-STEP states depending on which column contains the symbol. Although Labanotation symbols typically represent the direction of the toes, this implementation focuses solely on forward movement, disregarding direction, i.e., the shape of the symbols.

Among the possible states in the support columns, the condition where neither column contains a symbol is excluded for the current study. This condition represents either jumping with both feet off the ground or remaining stationary in a previous state, neither of which occur in the Japanese dance under consideration. As a result, only three states are defined and used in this context.

A lower-body task is defined as a movement that transitions to one of three states. As shown in Fig. 8.5b, tasks correspond to possible transitions among the states: STAND, L-STEP, R-STEP, and SQUAT. Beyond the support columns, the SQUAT state, described

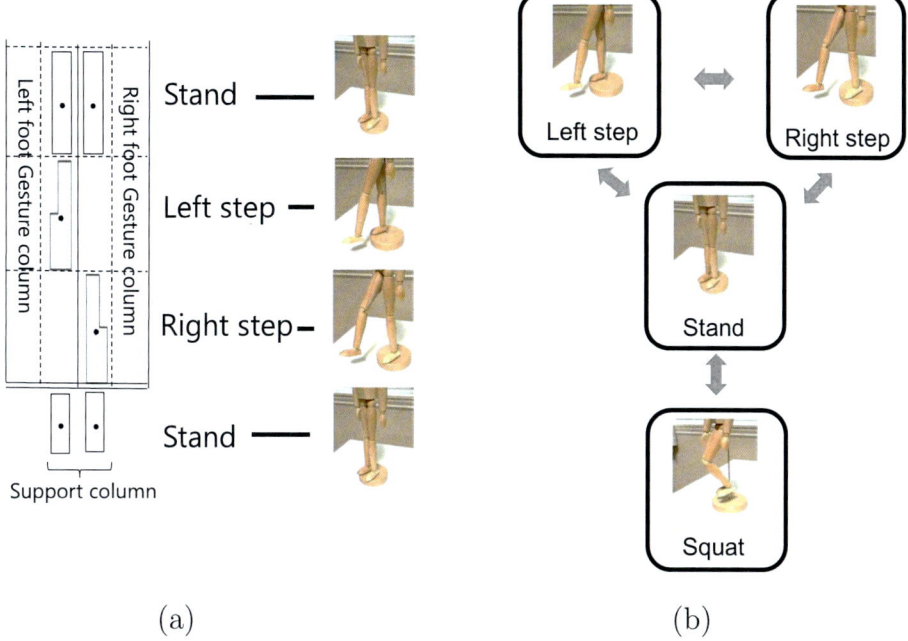

$$(a) \qquad\qquad\qquad (b)$$

**Fig. 8.5** Lower body Labanotation. **a** Lower body labanotation (support column only). **b** State transitions in lower body

in the waist column, is particularly crucial for reproducing the Japanese Aizu-bandaisan dance. This state is also incorporated into the lower-body states.

Table 8.1 presents the skill parameters associated with each task. In the current implementation, all tasks include start time ($t_s$) and end time ($t_e$) parameters to ensure synchronization with the music. This enables the duration of each task's movement to align with the music, with a sequence of leg movements represented as a task sequence organized along a timeline. By adhering to these timings, the rhythm of the movements is accurately conveyed.

As illustrated in Fig. 8.6, the positional parameters of a STEP task are defined using the relative coordinate system $\Sigma_{support}$, where the positive direction of the $z$-axis corresponds to the vertical upward direction. During the task, the support foot remains stationary on the floor, and the $z$-axis of $\Sigma_{support}$ is aligned with the $z$-axis of the world coordinate system, ensuring that the plane of the support foot matches the floor plane.

The end time of the task is represented as $t_e$, with $\mathbf{x}_e$ denoting the horizontal position where the swing leg's foot makes contact with the ground, and $\mathbf{r}_e$ representing the foot's orientation at this moment. Additionally, the position $\mathbf{x}_m$ and orientation $\mathbf{r}_m$ of the swing leg's foot at the midpoint $t_m$ of the STEP are defined as skill parameters. The trajectory of the swing leg's foot is designed to pass through this midpoint. Furthermore, the rotation

**Table 8.1** Skill parameters

| Skill | Notation | Skill parameters |
|---|---|---|
| Step | $t_s$ | Start timing |
| | $t_m$ | Mid-point timing |
| | $t_e$ | End timing |
| | $\mathbf{x}_m$ | Foot position at the mid-point timing |
| | $\mathbf{r}_m$ | Foot pose at the mid-point timing |
| | $\mathbf{x}_e$ | Foot position at the end timing |
| | $\mathbf{r}_e$ | Foot pose at the end timing |
| | $\theta_e$ | Body rotation at end timing |
| Squat | $t_s$ | Start timing |
| | $t_m$ | Mid-point timing |
| | $t_e$ | End timing |
| | $d_m$ | Squat depth at the mid-point |
| Stand | $t_s$ | Start timing |
| | $t_e$ | End timing |

**Fig. 8.6** Lower body coordinate system

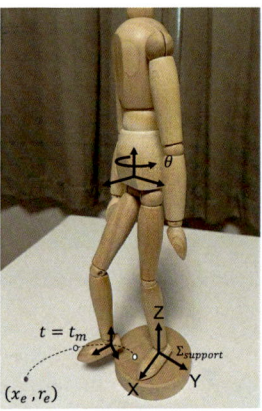

angle of the waist at the end time, $\theta_e$, represents movements involving waist rotations, such as those occurring during turns.

For a SQUAT task, the lowest vertical position of the waist is designated as the midpoint. The associated parameters include the time $t_m$ at this midpoint and the vertical difference $d_m$ from the starting position. These parameters are crucial for accurately representing and executing the SQUAT task, ensuring that its fundamental characteristics are preserved and effectively modeled.

For all tasks, the position and orientation at the start time, $t_s$, are not explicitly included as parameters. Instead, they inherit the execution results of the preceding task, aligning with the conventions of other LfO systems.

## 8.2.2 Lower-Body Task Recognition

Lower-body task recognition was conducted directly from motion capture data, without relying on Labanotation. This decision stemmed from the discrepancy between the gesture information provided by Labanotation and the parameters necessary for the robot's execution. The process involves first identifying the active time domain for each task, followed by extracting the positional and orientational parameters for each detected task.

### 8.2.2.1 Task Recognition

The recognition of tasks for the lower body relies directly on motion capture data while incorporating the concept of Labanotation, albeit without fully utilizing Labanotation itself. For upper body movements, a direct mapping from Labanotation to robotic motion is achievable, as the absence of environmental constraints enables straightforward imitation. In contrast, lower body movements present a unique challenge. Human soles are inherently soft, whereas the soles of robots used in implementation are rigid. This fundamental difference complicates the direct application of Labanotation to robot movements. Consequently, it becomes essential to perform task recognition tailored to the mechanical and structural attributes of robotic movements.

In the case of Labanotation for normal human walking, transitions are typically described from a STEP state to another, as illustrated in Fig. 8.5a. During the final phase of a step, when the swing foot contacts the ground, humans utilize the softness of their soles to facilitate weight transfer. Although the foot's overall position remains stationary during this period, the deformation of the sole plays a key role in completing the step. This phase is considered part of the STEP in Labanotation.

The rationale behind this lies in Labanotation's fundamental focus: describing the purpose of motion rather than the motion itself. In this specific case, the purpose is the forward movement of the sole, making the aforementioned time segment integral to achieving that objective.

For robots with rigid soles, weight transfer occurs exclusively through waist movement once both feet are in contact with the ground. Unlike humans, robots do not rely on sole deformation. Consequently, it is more practical to explicitly define this phase as a STAND state rather than merging it into the STEP state. This adjustment leads to the recognition of walking movements not as direct STEP-to-STEP transitions but as passing through the STAND state. The sequence thus becomes STEP, STAND, and STEP, as depicted in Fig. 8.7.

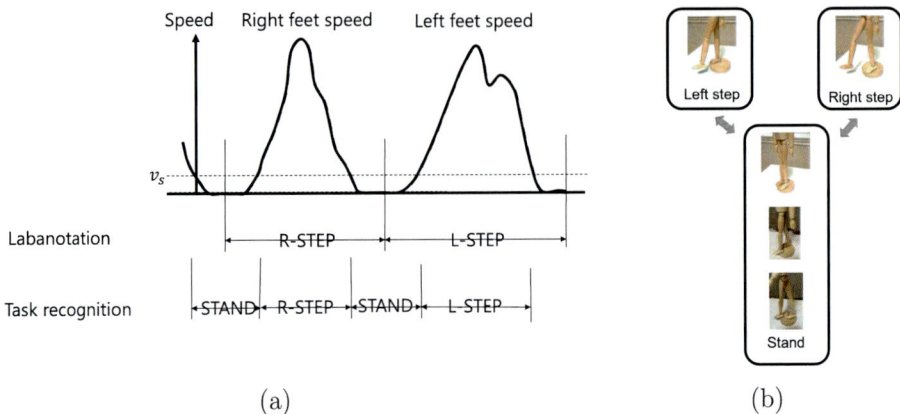

(a)                                                              (b)

**Fig. 8.7** Modification of the definition and transition of lower body states. **a** In the Labanotation, the state of both feet being in contact with the ground during the transition from one step to the next is treated as part of the step. However, in our current implementation, this hidden state is explicitly defined as an independent state. For its detection, a threshold value $v_s$ is used, where the STAND state is recognized when the foot speed drops below this threshold. **b** The transition does not proceed directly from one STPE state to the next one; instead, it passes through in an intermediate STAND state

This approach also simplifies the method of STEP recognition. A threshold value, $v_s$, is introduced for the swing leg's velocity. When the velocity falls below this threshold, the movement is classified as a STAND state. Conversely, movements exceeding the threshold velocity are identified as a STEP state. This classification streamlines the process and enhances efficiency in processing motion data. See Fig. 8.7a.

When identifying STEP tasks, the analysis of foot speed is prioritized. A segment is classified as a STEP task if the foot velocity exceeds a predefined threshold for a sustained period, and the distance traveled by the foot during this interval also surpasses a specified threshold. The longest segment meeting these criteria is then designated as the STEP task segment.

A STAND task is identified as the period during which neither a R-STEP nor L-STEP task is being performed, occurring after the detection of both R-STEP and L-STEP tasks. This classification is analogous to a hold task in upper-body movements.

To recognize a SQUAT task, the focus is placed on the trajectory of the vertical position of the waist. Assuming that the waist lowers and then returns once, the lowest point is identified as the midpoint. The recognition of SQUAT tasks is performed independently of the detection of STEP and STAND tasks.

## 8.2.2.2 Skill Parameter Extraction

For each task, the start time ($t_s$), end time ($t_e$), and midpoint time ($t_m$) are determined as the timing-related skill parameters. Additionally, positional and orientational skill parameters are derived from the marker positions at these specified times.

When collecting STEP skill parameters, it is essential to verify the validity of the midpoint. An interpolated trajectory is generated using a quadratic function based on the position and velocity of the swing foot at the start and end points of the STEP task. If a discrepancy exists between this interpolated trajectory and the actual trajectory, a midpoint is introduced. This midpoint is defined as the point where the difference between the interpolated trajectory and the actual trajectory is at its maximum.

The relative coordinate system $\Sigma_{support}$ is defined based on the position and orientation of the support leg's foot sole at the task's start time $t_s$. The $z$-axis of $\Sigma_{support}$ is aligned with the $z$-axis of the world coordinate system to ensure consistency. The position and orientation of the swing leg's foot sole at times $t_m$ (if the midpoint is valid) and $t_e$, as determined from markers, are transformed into coordinates within $\Sigma_{support}$ and utilized as skill parameter values. Likewise, the orientation of the waist is computed using data from several markers attached to it, and the Yaw angle $\theta_e$ is determined within the $\Sigma_{support}$ coordinate system.

For a SQUAT task, the standard vertical position of the waist is derived from data collected by waist markers at times $t_s$ and $t_e$. Assuming the waist lowers and then returns once during the movement, the lowest point in the trajectory is identified as the midpoint. Subsequently, the vertical difference $d_m$ between this midpoint and the standard vertical position is extracted as a skill parameter.

## 8.3   Runtime System

### 8.3.1   Dance Motion Generation

The system, as depicted in Fig. 8.8, comprises three primary components. First, the trajectory generator processes the sequence of Labanotations for the robot's upper and lower body, independently generating joint angle trajectories for each segment using designated agents [14]. These trajectories are then integrated and refined by dynamic filters to ensure stable lower body movements. Finally, the finer module ensures the overall coherence and consistency of the robot's full-body motion.

- **Trajectory generator** activates the agents corresponding to the input Labanotation. These agents generate joint trajectories by considering the skill parameters in the task sequence, along with factors such as target ZMP, waist position, and torso rotation.
- **Dynamic filter** refines the leg joint angle trajectories by incorporating upper body movements, providing a holistic approach to ensure stable and accurate robot motion. In particular, the ZMP and yaw compensation filters are applied: the ZMP compensation

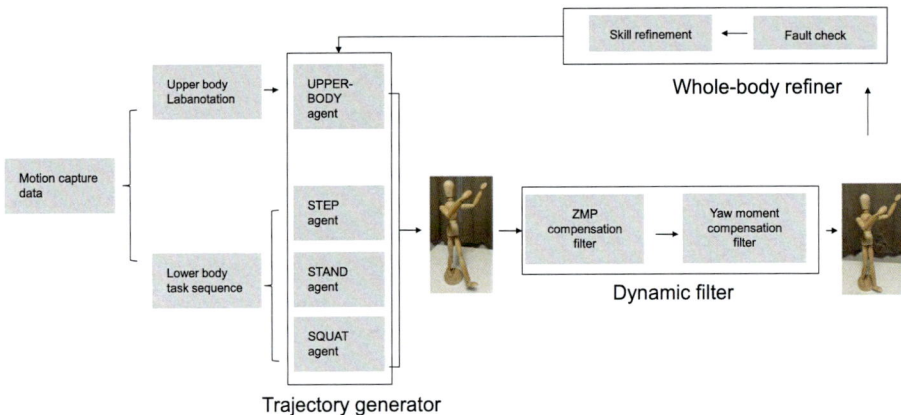

**Fig. 8.8** Dance motion generation system

filter prevents falls by maintaining balance, and the yaw-axis moment compensation filter corrects rotational forces to avoid spins caused by foot slippage. Together, these filters ensure controlled, smooth, and stable movements, preventing unexpected disruptions during performance.

- **Refiner** systematically examines the generated joint trajectories with meticulous attention to the overall body configuration and the robot's operational performance. Utilizing a simulator, it performs collision avoidance checks across various body segments and evaluates reachability constraints imposed by the robot's mechanisms. Based on these evaluations, the refiner adjusts and fine-tunes the movements to ensure seamless execution and alignment with the robot's capabilities.

These calculations are executed iteratively within a loop, adhering to a specified time resolution, $\Delta t$, as necessitated by the control system of the robot. For the HRP2 robot used in this context, the time resolution is set to 5 milliseconds. Within each iteration of the loop, the joint angles are systematically calculated.

## 8.3.2   Trajectory Generator

### 8.3.2.1 Upper-Body Agent

To calculate the upper body joint angles while ensuring mechanical feasibility and accurate motion representation, the system begins by deriving angles from Labanotation and connecting them using spline functions. These initial trajectories are then refined to meet the robot's specifications, specifically by addressing joint angle ranges and angular velocity constraints. For joint angle ranges, a mapping function comprising multiple polynomials applies local scaling in areas where original angles exceed limits, accounting for both time and values. Pollard's method [184] is employed to handle angular velocity constraints. These adjustments allow the trajectories to conform to joint-specific constraints while preserving the overall integrity of the motion, achieving a balance between precise depiction and operational limitations.

### 8.3.2.2 Lower-Body Agent

Lower-body agents operate within defined time frames, determined by their start and end times specified as skill parameters. Upon reaching the start time, the agent becomes active and begins generating the joint angle trajectories. Once the end time is reached, the agent ceases its operation. This structured timing ensures precise coordination and execution of the trajectories within the designated time intervals.

**STEP agent**: The STEP agent is responsible for calculating joint angles to manage the position and orientation of the swing leg's sole as the yaw-axis orientation of the waist, while preserving the position and orientation of the support leg's sole from its previous state. To determine the swing leg trajectory, $\mathbf{r}(t)$, it utilizes the midpoint and endpoint positions, $\mathbf{r}_m$ and $\mathbf{r}_e$, defined within the skill parameters. Initially, a cubic polynomial is employed to generate a smooth trajectory passing through the start position, midpoint, and endpoint, ensuring continuity and precision in motion.

In this initial trajectory, execution errors in the actual robot may cause a significant impact from the vertical reaction force when the foot sole contacts the floor, leading to instability in the robot's behavior. To mitigate this, an execution parameter $h_v$ is introduced, setting the velocity just before landing lower than that obtained from the original polynomial trajectory. This approach helps to reduce the impact force and improve the stability of the robot's movements, ensuring smoother and more reliable performance during task execution.

The final position of the waist is determined by the ZMP compensation filter. However, the STEP skill provides an initial input to this filter by estimating the trajectory of the waist position based on the start and end positions of the swing leg.

**STAND agent**: The STAND agent is to maintain the position and posture of the soles as it was at the beginning.

The primary role of the STAND agent is to plan the trajectory of the Zero Moment Point (ZMP) during the STAND task. ZMP is defined as the point where the resultant force of gravity and inertial forces intersects the ground. For a robot to maintain stability and avoid falling, the ZMP must reside within the convex hull of the support surface, such as the foot sole, as described in [185].

At the beginning of the STAND task, the ZMP is located within the convex hull of the supporting foot from the previous step. By the end of the STAND task, the ZMP must shift to the convex hull of the opposite supporting foot to ensure the next STEP task can proceed stably. For optimal stability in robotic motion, it is desirable for the ZMP to be positioned as centrally as possible within the convex hull. Considering these conditions, the STAND agent calculates the ZMP transfer trajectory while accounting for the preceding and subsequent tasks. This trajectory is referred to as the target ZMP trajectory.

The design policy for the target ZMP trajectory focuses on ensuring stability. Specifically, during the STAND task, the ZMP is maintained at its initial position for a fixed duration at the start, $t_z$. It then relocates to the central position of the convex hull encompassing both supporting feet. Then, it remains stationary for another fixed duration, $t_z$. This trajectory is approximated using a fifth-order polynomial.

This strategy accounts for stability by enabling the robot to dissipate disturbances from the previous step, execute unstable weight transitions, and pass through the most stable region of the ZMP before transitioning to the next position. After the ZMP movement stabilizes at the new position, the control is handed over to the next STEP agent. For further details, refer to [15].

**SQUAT agent**: The agent generates a waist trajectory using a cubic polynomial, ensuring it is in the standard position at the start and end times, and reaches the position specified by the skill parameter $d_m$ at the midpoint. In this implementation, the initial position of the waist is set as the standard position.

### 8.3.3  Dynamic Filters

#### 8.3.3.1 ZMP Compensation

To prevent the robot from falling, it is necessary to consider the overall balance of the body. Since the robot and the human body differ in terms of foot sole flexibility and weight distribution, it is not possible to solve the robot's balance problem using human motion trajectories. Instead, the robot needs to solve the problem independently according to its own body. The ZMP compensation filter adjusts the horizontal position of the robot's waist to maintain overall balance.

The robot's movements are generated under the assumption that the sole of the supporting leg is always in full contact with the floor. This contact condition ensures stability, as maintaining ZMP within the sole of the foot prevents the robot from falling. Mechanically, the ZMP condition corresponds to the calculated ZMP being located within the convex hull (supporting region) on the floor formed by the foot sole of the supporting leg(s). This convex hull defines the area within which the ZMP must remain to maintain balance.

To simply meet the ZMP condition, the ZMP can be positioned anywhere within the convex hull. However, for optimal stability, it is preferable to place the ZMP near the center of the convex hull. Additionally, to enable smooth and repetitive stepping motions, the ZMP trajectory should form a continuous and smooth curve between consecutive steps while satisfying the ZMP condition. Based on these considerations, the target ZMP trajectory is calculated by the STAND agent, ensuring both stability and fluid motion. Adjustments to the joint angles ensure that the calculated ZMP aligns with the target ZMP, resulting in a stable and balanced gait.

Several methods to achieve this target ZMP alignment have been proposed by researchers, including Nishiwaki et al. [186]. This implementation adopts Nishiwaki et al.'s method, which works by taking the target ZMP trajectory and the calculated ZMP trajectory from unmodified movements as inputs. It outputs an approximate correction for the horizontal position of the robot's waist to align the calculated ZMP with the target ZMP. By repeatedly applying this correction process, a solution with sufficient accuracy can be achieved.

The ZMP compensation filter primarily adjusts the horizontal position of the waist. First, for each $\Delta t$ frame, the joint angles and angular velocities of all joints, including those of the upper body, are prepared according to the link model. Then, forward kinematics calculations are performed, with the sole of the current supporting legs fixed to the ground as the origin. From the resulting values of the total center of mass, momentum, and angular momentum of the entire body, the ZMP is calculated.

The calculated ZMP trajectory obtained in this manner is input into the method proposed by Nishiwaki et al. [187], alongside the target ZMP trajectory generated by the STAND agent. Based on the output of this method, the horizontal component of the waist trajectory is adjusted. Once the adjustment amount is determined, inverse kinematics calculations between the sole and the waist are performed again to update the joint angle trajectories of the legs.

The ZMP compensation filter plays a crucial role in maintaining the timing, position, and orientation of the robot's leg movements. It ensures that the upper body's joint angle trajectories and skill parameters are adhered to, preserving key characteristics such as those required for tasks like dancing. While the filter operates effectively, the distinctive attributes of the motion are seamlessly maintained.

### 8.3.3.2 Yaw-Axis Moment Compensation Filter

If the Yaw axis moment exerted by the robot through the foot sole exceeds the moment due to friction between the foot sole and the floor, the foot sole may slip, causing the entire body to spin. Even if the original human movement does not involve spinning, there is a possibility that the generated robot movement may result in a spin.

Tamiya et al. [188] proposed a method to maintain the overall balance of a robot supported on one leg by compensating the entire body's moment and keeping it below a certain value. By incorporating the Yaw axis moment compensation part of this method as a filter, it generates motion trajectories that prevent spinning during execution.

### 8.3.4   Skill Refinement

Until now, the generation of the robot's movements has focused on contact conditions. However, apart from this condition, there may be obstacles that make the generated movements impossible for the robot to execute. Possible obstacles include deviation from the reachable range of the toes, occurrence of self-collision, exceeding joint angle range, or joint angular velocity limits. These obstacles arise because skill parameters obtained from human movements cannot be executed by the robot as is, due to differences in body shape and mechanical constraints of the robot. In such cases, it is necessary to adjust the skill parameters to match the robot's body.

Since the skill parameter values were originally executed by humans, even if obstacles occur with the robot, in most cases, the solution to avoid these obstacles can be found close to the original values. Additionally, the number of skill parameters is not very large, so the modification candidates are limited. Due to these characteristics, many obstacles can be resolved by relatively simple skill parameter adjustment rules.

## 8.4   Performance

This system uses motion capture to acquire human body movements. The optical system used here can capture the three-dimensional positions of 34 markers at a rate of 120 frames per second. The chosen dance for this study is the "Aizu Bandai-san Dance." Although the dance has a relatively gentle pace, it includes many characteristic movements involving the entire body. The dance was performed by an instructor to the same music, and the motion trajectories were captured as marker trajectory data. The first 35 s, consisting of four repeating patterns, were extracted and used as the data.

The HRP-2 [189], a bipedal humanoid robot with a full body and 30° of freedom in its joints, was employed for the performance. It closely resembles human proportions, standing at a height of 1.54 m and weighing 56 kg.

The standard height of the waist (the vertical position of the waist link origin), $h_w$, is a crucial parameter related to the reachable range of the foot sole when it is in contact with the floor. This height should be set slightly lower than the maximum height the waist can achieve while maintaining foot sole contact with the floor. Reducing $h_w$ increases the reachable range of the foot sole but may result in an unnatural posture with deeply bent knees. Taking these factors into account, $h_w$ is set to 0.61 m, compared to the maximum possible value of 0.71 m for HRP-2.

The standard height of the step, $h_s$, and the height to deceleration, $h_v$, are set to 0.05 and 0.005 m, respectively. The control loop duration, $\Delta t$, and the timing to maintain ZMP, $t_z$, are set to 5[ms] and 25 ms, respectively.

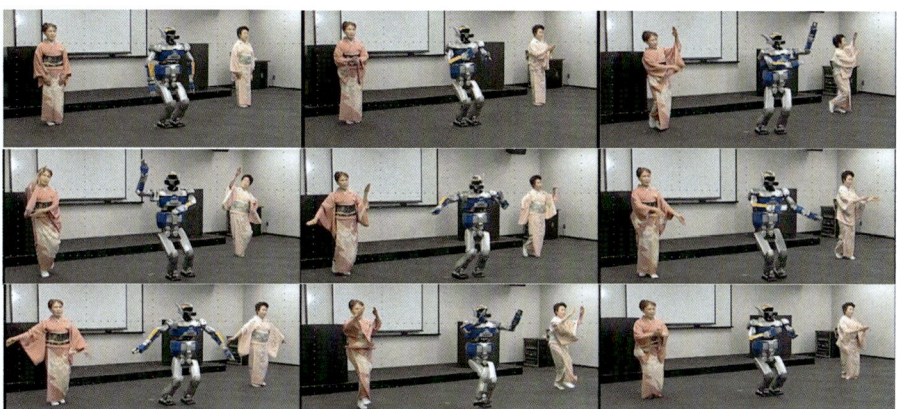

**Fig. 8.9** Co-dancing with Humanoid HRP-2 and human dancers

Figure 8.9 depicts the final realization of the Aizu Bandai-san Dance performance by HRP-2 and the instructors, conducted in 2003 at the Institute of Industrial Science, the University of Tokyo. You can watch the robot dance on YouTube.[2]

---

[2] https://youtu.be/PGKGXxwp6LM.

# Epilogue

## Piaget Theory and LfO

The design philosophy of "Learning-from-Observation (LfO)" draws inspiration from Piaget's theory [6] of cognitive development in children. Piaget's theory delineates four distinct stages of development: the sensorimotor stage (0–2 years), the preoperational stage (2–7 years), the concrete operational stage (7–11 years), and the formal operational stage (11 years and beyond). During the sensorimotor stage, which spans from birth to approximately two years of age, children exhibit behaviors such as repetitive actions, referred to as counter actions. Subsequently, imitation behaviors, wherein children mimic the actions of their parents, begin to manifest. The concept of Learning-from-Observation is fundamentally derived from these imitation behaviors in the sensorimotor stage (around 1.5 years).

From a robotics perspective, the counter actions observed during the sensorimotor stage can be likened to reinforcement learning used for hand-eye calibration and the development of skill agents. In this period, children are known to engage in what Piaget refers to as "mini-experiments," where they repeatedly perform similar actions, but with slight variations in position or angle. This can be compared to the process of randomization in reinforcement learning, where skill agents explore different actions to optimize their performance.

During the sensorimotor stage, it is observed that following the emergence of counter actions, imitation behaviors, where children mimic their parent's actions, also manifest. When considering this from a robotics perspective, it can be understood that infants generate their physical behaviors by obtaining an overall plan based on visual information and utilizing the skill agents acquired previously through counter actions. This aligns well with the concept of LfO.

K. Ikeuchi et al., *Learning-from-Observation 2.0*, Synthesis Lectures on Computer Vision, https://doi.org/10.1007/978-3-032-03445-8

It is imperative to note that while the visual space is shared between the infant and the mother, the force sensory space is not. Consequently, although an infant can visually mimic the parent's actions at a global level, the refinement of actions using force sensations is conducted independently on the infant's hardware, without direct instructional data from the mother. The feedback from force sensations is employed in the skill agents that were developed through counter actions during the earlier stage.

In LfO, instead of utilizing force information obtained directly through teleoperation as teaching data, as seen in imitation learning or programming-by-demonstration, force information is employed as localized feedback intrinsic to the hardware. Consequently, the system is primarily designed with visual information serving as the main source for teaching.

## Top Down Approach

LfO can be regarded as a form of visual recognition predicated on pre-existing frameworks, drawing inspiration from Minsky's frame theory [5]. In contrast to methodologies such as "Programming-by-Demonstration" or "Learning-from-Demonstration," which directly replicate the demonstrated trajectories, LfO interprets these trajectories through abstract structures known as task models, which are designed in a top-down manner using robotics theories.

Human actions, particularly the demonstrated trajectories of hand positions, are segmented into meaningful intervals based on visual and/or verbal input. Each interval is then tagged with a "what-to-do" label among the possible labels (in the skill library), indicating the current action. Corresponding to this labeling, a task model, which is a Minsky frame representation, is activated to interpret each interval.

Each task model delineates in what way to characterize/interpret localized trajectories, thereby facilitating the extraction of features that define a particular task. These characteristics are subsequently stored as skill parameters, which can be considered "where-to-do" parameters, for guiding the execution of the task. From a computer science perspective, it is useful to conceptualize a task model as an operator and the skill parameters as its operands. Just as the types of operands vary with the operator, the types of parameters required differ depending on each task model.

The majority of skill parameters are not the trajectories themselves but rather a collection of localized vectors gathered according to the task requirements. These vectors are typically represented in an object-centered coordinate system to ensure reusability. For instance, in a grasping task, the vector indicates the optimal approach direction on the coordinate system of the target object. Similarly, in a placement task, it specifies the best direction to place the object on a table, using the local coordinate system of the target table. These vectors represent localized motion information relevant to specific tasks.

Adopting this localized description approach offers several advantages over other similar methods.

- **Reusability**: One well-known drawback of direct trajectory learning methods, such as Programming-by-Demonstration (Pd) or Learning-from-Demonstration (LfD), is the difficulty in reusing trajectory information when object positions change. In contrast, Learning-from-Observation (LfO) utilizes object-centered local information exclusively, enhancing the potential for reusability when object positions change.
- **Reduced Fatigue**: In Pd and LfD, the system mimics the entire trajectory, requiring the demonstrator to pay attention to every aspect of the demonstration. This can lead to significant fatigue for the demonstrator. LfO, on the other hand, collects information only from critical moments, allowing the demonstrator to focus on these key aspects, thereby reducing fatigue.
- **Error correction**: The system can filter out observational noise by utilizing constraints derived from the task model [10].

These advantages underscore the value of employing localized descriptions in Learning-from-Observation.

For each task model, a dedicated skill agent is prepared to execute the task. These skill agents essentially embody the "how-to-do" aspect. They are pre-trained using reinforcement learning, with reward functions based on top-down knowledge, primarily state transitions, to align with each robot's specific hardware. These agents generate the necessary movements to accomplish tasks based on the policies with the skill parameters provided by the task models as well as force feedback obtained locally in real-time.

Since these skill agents are tailored to specific hardware, any hardware changes necessitate only the replacement of these skill agents without altering the upstream task recognition components. This allows the system to be executed using the same observational framework across different execution hardware.

## Two-Stage Approach

Recently, there has been a growing trend towards developing end-to-end (E2E) systems utilizing foundation models powered by Vision Language Models (VLMs). Given the established advantages of LLMs in the language and vision domains, their application in robotics holds substantial potential for further advancements.

On the other hand, the difficulty of the E2E approach lies in the need for hardware-specific training data. Previously, the output of LLMs was primarily directed at humans with consistent performance, eliminating the need to prepare training data based on user response. However, with the E2E approach, LLMs directly control the robot's hardware, necessitating hardware-specific training data. When the robot's hardware changes, it becomes essential to retrain all foundation models from upstream to downstream.

Furthermore, the tasks required for robots, as outlined in this book, range from simple pick-and-place operations to wiping tasks defined by semantic constraints, as well as assembly tasks for mechanical components such as screw tightening, which require visual feedback. Additionally, the manipulation of flexible objects, such as tying a knot, is also included. To train an E2E foundation model capable of adapting all these tasks to changes in the upstream task sequence, an astronomical amount of data will likely be required.

LfO2.0, introduced in this book, adopts a two-stage approach to prevent information overload. It assigns only high-level task recognition to the LLM, while execution is managed by hardware-dependent skill agents. The upstream learning can be conducted using general vision-language data; in fact, LfO2.0 utilizes off-the-shelf LLMs. Meanwhile, the scope of downstream skills is clearly defined, allowing reinforcement learning to efficiently train within those boundaries. When hardware changes, only the downstream skill agents need to be retrained, leaving the upstream LLM untouched. When a specialized skill is required, only that specific skill needs to be trained and then added to the skill library.

The two-stage approach is illustrated using the example of driving a car, where each skill corresponds to components such as the brake pedal and steering wheel. See Fig. e.1. Current E2E approaches attempt to handle everything from route planning and situational assessment to engine and brake control in a unified manner. In contrast, the two-stage approach described in this book defines a skill set consisting of the brake pedal, accelerator pedal, and steering wheel. The upstream LLM is responsible for planning and situational assessment, including actions like pressing the accelerator pedal or steering. The corresponding skill agents then execute acceleration or turning movements. By adopting the two-stage approach, even if the vehicle changes, there is no need to relearn planning and situational assessment for driving.

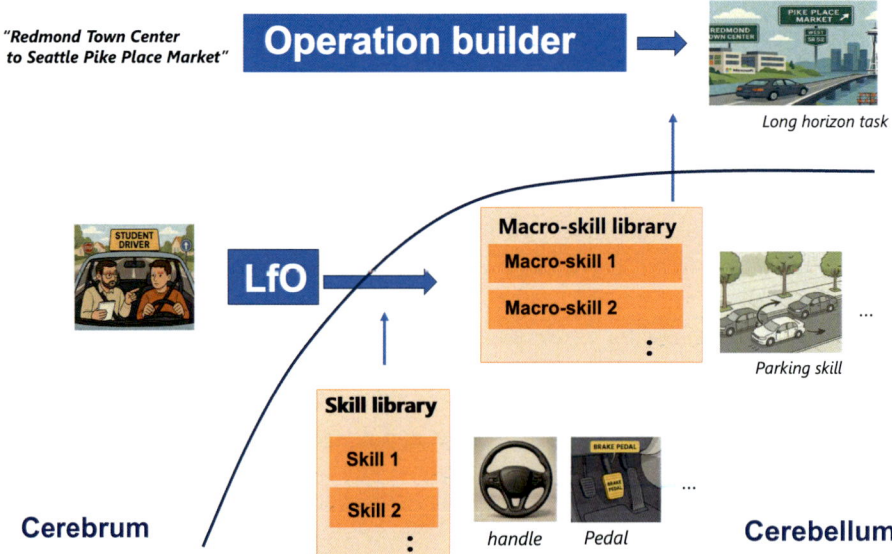

**Fig. e.1** Two-stage approach

Although not covered in this book, the two-stage approach is also effective for planning long-horizon tasks [190]. As task sequences grow in complexity and length, collecting sufficient training data for end-to-end (E2E) models becomes increasingly difficult. In contrast, the two-stage approach facilitates an inductive strategy. First, frequently utilized task groups essential for long-horizon tasks are pre-defined in LfO as Macro skills. Subsequently, an LLM decomposes long-horizon tasks into sequences of these Macro skills. This LLM, referred to as the Operation Builder, enables the mapping of the long-horizon task to executable robotic actions. The identification of optimal Macro skills for such tasks remains an open research question.

Drawing a parallel to neuroscience, one might consider the human brain's structure: the cerebrum and the cerebellum. The upstream task recognition can be likened to the role of the cerebrum, while the skill agents correspond to the role of the cerebellum. Referring back to Piaget's development theory, the repetition in the encounter behavior can be seen as the training of the cerebellar agents, while the later mimicry behavior can be interpreted as the cerebrum summarizing these agents. This analogy underscores the modular design, where the cerebrum (upstream task recognition) coordinates and integrates the cerebellar (skill agents) activities.

## Generality of Intermediate Representation

LfO2.0 is structured into two modules: the upstream portion, which obtains the task model, and the downstream portion, which executes the task model using a set of skill agents. For this segmentation to be meaningful and effective, it is essential to ensure that the types of tasks connecting the upstream and downstream modules possess generality. LfO 2.0 classifies tasks using three distinct theories to ensure generality: Labanotation, Closure Theory, and the Kuhn-Tucker Theory. These theories provide a robust framework for categorizing tasks, ensuring consistency and applicability across various scenarios.

LfO 2.0 employs a comprehensive approach to classifying human tasks based on the presence of contact, ensuring precise task categorization. Initially, tasks are broadly divided into two categories: those involving contact with the environment and those that do not. For tasks that do not involve contact, Labanotation is employed for classification.

For tasks involving contact, human actions are further subdivided into grasping actions, where the body transitions from a non-contact state to establishing contact with an object, and manipulative actions, where the body transitions between the contact states of the object with respect to the environment. Grasping tasks are classified using Closure Theory, while manipulative tasks are classified using the Kuhn-Tucker Theory.

**Generality of whole body representation** Let's consider the task classification in dance based on Labanotation. Labanotation involves sampling a series of postures (key poses) at specific time intervals (brief stop timings) in spatial terms. However, it does not record the trajectory between these key poses. According to the Gestalt view of movement, humans perceive dance as a continuous sequence of key poses. This tendency was observed in Perera's study [23], which found that humans focus only on the sequence of significant poses at critical moments. Thus, these perspectives suggest that when generating robot movements from human demonstrations, it is not necessary to imitate the entire spatiotemporal trajectory. Instead, a Labanotation-like approach that mimics only the key poses may be more effective.

By concentrating solely on reproducing postures from Labanotation, it is possible to overcome hardware challenges. Normally, differences in hardware between the demonstrating human and the robot—such as weight, height, and arm length—make it impossible for a robot to mimic all human actions accurately. This issue also arises, though to a lesser degree, between a human teacher and student. As a result, Labanotation expresses only the postures that appear similar to human spectators, without requiring the replication of precise trajectories.

In robot dances uploaded on YouTube[1] described in Chap. 8, only the key poses of the robot are designed to imitate those of a dance master. However, the trajectories of each part of the robot between key poses are generated to create stable postures to maintain balance and prevent falling. Consequently, the fine trajectories between the robot's key poses may differ from those of Master Yamada. Despite these differences, spectators barely notice any discrepancies and consider both to be performing the same dance when watching the video. This observation provides justification for using Labanotation, which expresses only key poses, as a method for classifying body movements in non-contact situations.

**Generality of grasp representation**   Grasping can be defined as the movement of the hand to establish a specific relationship with an object. Various classifications of grasping have been proposed for the grasping shapes of a hand, focusing on how the fingers of the human hand are spatially distributed in relation to the object. In contrast, the shape, number, and dimensions of a robot's fingers differ significantly from those of a human hand. Consequently, it is not possible to replicate the exact grasping shape of a human hand with a given robot hand. Therefore, we adopt a strategy to classify grasping motions according to the purpose of grasping, i.e., the degree of freedom of the object after it has been grasped.

Regarding the degrees of freedom of grasped objects, Yoshikawa proposed The Closure theory [17]. According to this theory, there are three types of closures defined based on the degrees of freedom of a grasped object: Passive form, which fixes only the position of the object; Passive force, which constrains all degrees of freedom of the object; and Active force, which constrains only some degrees of freedom of the object. By examining how each finger distribution grants degrees of freedom based on easily observable shapes, these shape distributions are classified into one of the three types.

---

[1] https://youtu.be/PGKGXxwp6LM.

For implementation, grasp motions that can achieve these closures using a given robot hand are pre-developed as part of a "grasp skill library" specific to the robot hardware. During execution, the observed grasping form is identified with the corresponding closure, and the grasp skill associated with this closure is executed. Consequently, the classification of grasping motions is also based on the Closure theory and can be considered sufficiently general.

**Generality of manipulation representation** Manipulation motion is defined as the operation of transitioning the contact relationship between the grasped object and the environment. In this case, the object has already been grasped by the robot and can be considered part of the robot's body. In fact, there are psychological reports that recognize a grasped tool as part of the body. Therefore, the constraints on the movement of the object are considered the same as those on the movement of the hand. The range of motion of a grasped object, in other words, the hand, can be expressed as the solution space of linear inequalities representing constraints from the environment. According to the Kuhn-Tucker theory, the topology of the solution space of these simultaneous linear inequalities is ten, and, thus, the number of contact states is also ten. Given that the upper bound of possible states is ten, the upper bound of state transitions among them, and consequently the number of manipulation tasks, is one hundred. In LfO2.0, a practical manipulation motion set is defined by incorporating only those transitions commonly observed in daily life into a library, thereby ensuring both generality and practicality.

As discussed above, we utilized Labanotation to classify actions in the absence of contact, Closure theory to classify grasping actions until contact occurs, and Kuhn-Tucker theory to classify manipulation actions, after contact, that facilitate transitions in the contact state. It is intriguing to inquire whether the intermediate representations automatically generated from data using the foundation model align with the intermediate representations constructed using these robotics theories.

## 90%AI

LfO is a teaching system based on Reddy's AI principle. In his Turing award lecture, Professor Reddy asserted that AI systems should be designed to manage 90% of straightforward and routine tasks, reserving the remaining 10% of exceptional and complex tasks for human intervention [191]. In our LfO, we also adhere to this principle, deriving insights from human demonstrations to resolve grasping strategy decisions and collision avoidance path determinations according to target objects and environments. For instance, concerning grasping, even for the same target object, the grasping strategy can vary depending on the intended subsequent action. Indeed, since the early stages of grasping research, there have been attempted to automatically generate grasping strategies based on the task's objectives.

However, the history of robotics research has shown that these automatic generation methods are challenging. Therefore, we derive these strategies from human demonstrations.

Collision avoidance is not automatically generated but derived from human input. In the example discussed in Sect. 2.3.2, critical waypoints for collision avoidance are deliberately demonstrated, such as moving the object outside the shelf to a distance where a collision can be averted before adjusting the height. Based on this, instead of generating a single `Bring` (PTG12) task, three `BringCarefully` (STG12) tasks are created. The human demonstrator resolves collision avoidance and provides global guidance for a collision-free path to the system, while the system, in turn, automatically generates the trajectory passing through these waypoints according to the global guidance.

This 90% AI concept is also applied in Labanotation-based Inverse Kinematics (IK), where the system solves the redundant degrees of IK starting from the initial posture provided by Labanotation. Humanoid robots typically exhibit a high degree of redundancy due to their numerous joints. Solving IK for such highly redundant systems is generally challenging. Furthermore, even when a solution is found, multiple solutions exist, and determining the optimal one depends on the task flow. Consequently, we employ a method that derives posture hints from human demonstrations and solves IK based on these hints.

These three examples demonstrate how the 90% principle facilitates reaching a global solution in problems with multiple local solutions by utilizing human demonstrations as the initial solution. The 90% AI approach signifies designing systems as human assistants, not as masters of the world, always considering the respective roles of humans and AI systems within a coexisting environment.

As Prof. Reddy points out, however, it is important not to stop at achieving a 90% system. Instead, after implementing the 90% AI system, it is essential to iteratively subdivide the remaining 10% into a 9:1 ratio, thereby advancing to the next stage's 90% AI system. Repeating this process will asymptotically approach a 100% AI system, but it is crucial never to surpass human capabilities.

This concept of 90% AI also informs future design principles for robotics. Specifically, it underscores the imperative that humans must always remain in control, with robots serving in a subordinate role. Robots should not evolve into *autonomous embodied AI* capable of physically overpowering humans or becoming adversarial entities, as portrayed in fictional narratives such as "The Terminator". Instead, the goal should be to develop *augmented embodied AI* that consistently acts as an ally to humans and provides meaningful assistance.

**Fig. e.2** Enemy or ally

A compelling model for this vision can be found in the Japanese anime "Doraemon", where the robot character exemplifies unwavering support and companionship. With this perspective, we conclude this book (Fig. e.2).

Katsushi Ikeuchi
Naoki Wake
Jun Takamatsu
Kazuhiro Sasabuchi

*February 2025*
*On a rare snowy day*
*In Redmond*

# References

1. Y. Kuniyoshi, M. Inaba, H. Inoue, Learning by watching: extracting reusable task knowledge from visual observation of human performance. IEEE Trans. Robot. Autom. **10**(6), 799–822 (1994)
2. S. Schaal, Is imitation learning the route to humanoid robots? Trends Cogn. Sci. **3**(6), 233–242 (1999)
3. A. Billard, S. Calinon, R. Dillmann, S. Schaal, Robot programming by demonstration, in *Springer Handbook of Robotics* (2008), pp. 1371–1394
4. B.D. Argall, S. Chernova, M. Veloso, B. Browning, A survey of robot learning from demonstration. Robot. Auton. Syst. **57**(5), 469–483 (2009)
5. M. Minsky, *Society of Mind* (Simon and Schuster, 1988)
6. J. Piaget, *La Psychologie de l'intelligence* (Dunod, 2020)
7. K. Ikeuchi, R. Reddy, Learning from observation, in *Annual Research Review, CMU-RI* (1991)
8. R. Reddy, Three open problems in AI. J. ACM (JACM) **50**(1), 83–86 (2003)
9. K. Ikeuchi, T. Suehiro, Toward an assembly plan from observation. I. Task recognition with polyhedral objects. IEEE Trans. Robot. Autom. **10**(3), 368–385 (1994)
10. T. Suehiro, K. Ikeuchi, Towards an assembly plan from observation: Part II: correction of motion parameters based on fact contact constraints, in *Proceedings of the IEEE/RSJ International Conference on Intelligent Robots and Systems*, vol. 3 (IEEE, 1992), pp. 2095–2102
11. J. Takamatsu, T. Morita, K. Ogawara, H. Kimura, K. Ikeuchi, Representation for knot-tying tasks. IEEE Trans. Rob. **22**(1), 65–78 (2006)
12. A. Hutchinson-Guest, *Labanotation: The System of Analyzing and Recording Movement* (Theatre Arts Books, New York, 1970)
13. K. Ikeuchi, Z. Ma, Z. Yan, S. Kudoh, M. Nakamura, Describing upper-body motions based on Labanotation for learning-from-observation robots. Int. J. Comput. Vis. (IJCV) **126**(12), 1415–1429 (2018)
14. S. Nakaoka, A. Nakazawa, K. Yokoi, H. Hirukawa, K. Ikeuchi, Generating whole body motions for a biped humanoid robot from captured human dances, in *2003 IEEE International Conference on Robotics and Automation (Cat. No. 03CH37422)*, vol. 3 (IEEE, 2003), pp. 3905–3910

K. Ikeuchi et al., *Learning-from-Observation 2.0*, Synthesis Lectures on Computer Vision,
https://doi.org/10.1007/978-3-032-03445-8

15. S. Nakaoka, A. Nakazawa, F. Kanehiro, K. Kaneko, M. Morisawa, H. Hirukawa, K. Ikeuchi, Learning from observation paradigm: leg task models for enabling a biped humanoid robot to imitate human dances. Int. J. Robot. Res. **26**(8), 829–844 (2007)

16. N. Wake, A. Kanehira, K. Sasabuchi, J. Takamatsu, K. Ikeuchi, Gpt-4v (ision) for robotics: multimodal task planning from human demonstration, in *IEEE Robotics and Automation Letters* (2024)

17. T. Yoshikawa, Passive and active closures by constraining mechanisms. J. Dyn. Syst. Meas. Contr. **121**(3), 418–424 (1999)

18. H. Kuhn, A. Tucker, Linear inequalities and related systems. Bull. Am. Math. Soc **63**, 202–203 (1957)

19. N. Wake, I. Yanokura, K. Sasabuchi, K. Ikeuchi, Verbal focus-of-attention system for learning-from-observation, in *2021 IEEE International Conference on Robotics and Automation (ICRA)* (IEEE, 2021), pp. 10 377–10 384

20. N. Wake, R. Arakawa, I. Yanokura, T. Kiyokawa, K. Sasabuchi, J. Takamatsu, K. Ikeuchi, A learning-from-observation framework: One-shot robot teaching for grasp-manipulation-release household operations, in *IEEE/SICE International Symposium on System Integration (SII)* (IEEE, 2021), pp. 461–466

21. X. Zhou, R. Girdhar, A. Joulin, P. Krähenbühl, I. Misra, Detecting twenty-thousand classes using image-level supervision, in *European Conference on Computer Vision* (Springer, Berlin, 2022), pp. 350–368

22. I. Yanaokura, N. Wake, K. Sasabuchi, R. Arakawa, K. Okada, J. Takamatsu, M. Inaba, K. Ikeuchi, A multimodal learning-from-observation towards all-at-once robot teaching using task cohesion, in *IEEE/SICE International Symposium on System Integration (SII)* (IEEE, 2022), pp. 367–374

23. M. Perera, S. Kudoh, K. Ikeuchi, Keypose and style analysis based on low-dimensional representation. J. Inf. Process. Soc. Jpn. **50**, 1234–1249 (2009)

24. G.A. Miller, The magical number seven, plus or minus two: some limits on our capacity for processing information. Psychol. Rev. **63**(2), 81 (1956)

25. T. Winograd, Understanding natural language. Cogn. Psychol. **3**(1), 1–191 (1972)

26. P. Pramanick, H.B. Barua, C. Sarkar, Decomplex: task planning from complex natural instructions by a collocating robot, in *2020 IEEE/RSJ International Conference on Intelligent Robots and Systems (IROS)* (IEEE, 2020), pp. 6894–6901

27. S.G. Venkatesh, R. Upadrashta, B. Amrutur, Translating natural language instructions to computer programs for robot manipulation, in *2021 IEEE/RSJ International Conference on Intelligent Robots and Systems (IROS)* (IEEE, 2021), pp. 1919–1926

28. Y. Jiang, A. Gupta, Z. Zhang, G. Wang, Y. Dou, Y. Chen, L. Fei-Fei, A. Anandkumar, Y. Zhu, L. Fan, VIMA: general robot manipulation with multimodal prompts **2**(3), 6 (2022). arXiv:2210.03094

29. M. Shridhar, L. Manuelli, D. Fox, Perceiver-actor: a multi-task transformer for robotic manipulation, in *Conference on Robot Learning* (PMLR, 2023), pp. 785–799

30. A. Brohan, Y. Chebotar, C. Finn, K. Hausman, A. Herzog, D. Ho, J. Ibarz, A. Irpan, E. Jang, R. Julian, et al., Do as I can, not as I say: grounding language in robotic affordances, in *Conference on Robot Learning* (PMLR, 2023), pp. 287–318

31. W. Huang, F. Xia, T. Xiao, H. Chan, J. Liang, P. Florence, A. Zeng, J. Tompson, I. Mordatch, Y. Chebotar, et al., Inner monologue: embodied reasoning through planning with language models (2022). arXiv:2207.05608

32. Y. Ding, X. Zhang, C. Paxton, S. Zhang, Task and motion planning with large language models for object rearrangement, in *2023 IEEE/RSJ International Conference on Intelligent Robots and Systems (IROS)* (IEEE, 2023), pp. 2086–2092

33. I. Singh, V. Blukis, A. Mousavian, A. Goyal, D. Xu, J. Tremblay, D. Fox, J. Thomason, A. Garg, ProgPrompt: generating situated robot task plans using large language models, in *2023 IEEE International Conference on Robotics and Automation (ICRA)* (IEEE, 2023), pp. 11 523–11 530

34. K. Namasivayam, H. Singh, V. Bindal, A. Tuli, V. Agrawal, R. Jain, P. Singla, R. Paul, Learning neuro-symbolic programs for language guided robot manipulation, in *2023 IEEE International Conference on Robotics and Automation (ICRA)* (IEEE, 2023), pp. 7973–7980

35. Z. Zhao, W.S. Lee, D. Hsu, Differentiable parsing and visual grounding of natural language instructions for object placement, in *2023 IEEE International Conference on Robotics and Automation (ICRA)* (IEEE, 2023), pp. 11 546–11 553

36. Y. Ding, X. Zhang, S. Amiri, N. Cao, H. Yang, C. Esselink, S. Zhang, Robot task planning and situation handling in open worlds (2022). arXiv:2210.01287

37. A. Zeng, M. Attarian, B. Ichter, K. Choromanski, A. Wong, S. Welker, F. Tombari, A. Purohit, M. Ryoo, V. Sindhwani, et al., Socratic models: composing zero-shot multimodal reasoning with language (2022). arXiv:2204.00598

38. J. Liang, W. Huang, F. Xia, P. Xu, K. Hausman, B. Ichter, P. Florence, A. Zeng, Code as policies: language model programs for embodied control, in *2023 IEEE International Conference on Robotics and Automation (ICRA)* (IEEE, 2023), pp. 9493–9500

39. S.S. Raman, V. Cohen, E. Rosen, I. Idrees, D. Paulius, S. Tellex, Planning with large language models via corrective re-prompting, in *NeurIPS 2022 Foundation Models for Decision Making Workshop* (2022)

40. Y. Xie, C. Yu, T. Zhu, J. Bai, Z. Gong, H. Soh, Translating natural language to planning goals with large-language models (2023). arXiv:2302.05128

41. A.K. Kovalev, A.I. Panov, Application of pretrained large language models in embodied artificial intelligence, in *Doklady Mathematics*, vol. 106, no. Suppl. 1 (Springer, Berlin, 2022), pp. S85–S90

42. M.A. Khan, M. Kenney, J. Painter, D. Kamale, R. Batista-Navarro, A. Ghalamzan-E, Natural language robot programming: NLP integrated with autonomous robotic grasping (2023). arXiv:2304.02993

43. F. Kaynar, S. Rajagopalan, S. Zhou, E. Steinbach, Remote task-oriented grasp area teaching by non-experts through interactive segmentation and few-shot learning (2023). arXiv:2303.10195

44. M. Skreta, N. Yoshikawa, S. Arellano-Rubach, Z. Ji, L. B. Kristensen, K. Darvish, A. Aspuru-Guzik, F. Shkurti, A. Garg, Errors are useful prompts: instruction guided task programming with verifier-assisted iterative prompting (2023). arXiv:2303.14100

45. W. Huang, P. Abbeel, D. Pathak, I. Mordatch, Language models as zero-shot planners: extracting actionable knowledge for embodied agents, in *International Conference on Machine Learning* (PMLR, 2022), pp. 9118–9147

46. J.J. Gibson, *The Ecological Approach to Visual Perception: Classic Edition* (Psychology Press, 2014)

47. J.K. Tsotsos, Active recognition, in *Computer Vision: A Reference Guide* (Springer, Berlin, 2021), pp. 15–23

48. K. Ikeuchi, M. Hebert, Task-oriented vision, in *Exploratory Vision: The Active Eye* (Springer, Berlin, 1996), pp. 257–277

49. A. Aydemir, A. Pronobis, M. Göbelbecker, P. Jensfelt, Active visual object search in unknown environments using uncertain semantics. IEEE Trans. Rob. **29**(4), 986–1002 (2013)

50. S. Baluja, D. Pomerleau, Dynamic relevance: vision-based focus of attention using artificial neural networks. Artif. Intell. **97**(1–2), 381–395 (1997)

51. J.K. Tsotsos, S.M. Culhane, W.Y.K. Wai, Y. Lai, N. Davis, F. Nuflo, Modeling visual attention via selective tuning. Artif. Intell. **78**(1–2), 507–545 (1995)

52. T. Nguyen, M. Dax, C.K. Mummadi, N. Ngo, T.H.P. Nguyen, Z. Lou, T. Brox, DeepUSPS: deep robust unsupervised saliency prediction via self-supervision, in *Advances in Neural Information Processing Systems*, vol. 32 (2019)

53. M. Cornia, L. Baraldi, G. Serra, R. Cucchiara, Predicting human eye fixations via an LSTM-based saliency attentive model. IEEE Trans. Image Process. **27**(10), 5142–5154 (2018)

54. E. Kazakos, A. Nagrani, A. Zisserman, D. Damen, Epic-fusion: audio-visual temporal binding for egocentric action recognition, in *Proceedings of the IEEE/CVF International Conference on Computer Vision* (2019), pp. 5492–5501

55. K. Sasabuchi, N. Wake, K. Ikeuchi, Task-oriented motion mapping on robots of various configuration using body role division. IEEE Robot. Autom. Lett. **6**(2), 413–420 (2021)

56. J. Jaroslavceva, N. Wake, K. Sasabuchi, K. Ikeuchi, Robot ego-noise suppression with labanotation-template subtraction. IEEJ Trans. Electr. Electron. Eng. **17**(3), 407–415 (2022)

57. K. Sasabuchi, D. Saito, A. Kanehira, N. Wake, J. Takamatsu, K. Ikeuchi, Task-sequencing simulator: integrated machine learning to execution simulation for robot manipulation (2023). arXiv:2301.01382

58. E. Coumans, Y. Bai, Pybullet, a python module for physics simulation for games, robotics and machine learning (2016–2021). http://pybullet.org

59. J. Schulman, F. Wolski, P. Dhariwal, A. Radford, O. Klimov, Proximal policy optimization algorithms (2017). arXiv:1707.06347

60. J. Collins, S. Chand, A. Vanderkop, D. Howard, A review of physics simulators for robotic applications. IEEE Access **9**, 51 416–51 431 (2021)

61. W. Zhao, J.P. Queralta, T. Westerlund, Sim-to-real transfer in deep reinforcement learning for robotics: a survey, in *IEEE Symposium Series on Computational Intelligence (SSCI)* (IEEE, 2020), pp. 737–744

62. N. Koenig, A. Howard, Design and use paradigms for gazebo, an open-source multi-robot simulator, in *IEEE/RSJ International Conference on Intelligent Robots and Systems (IROS) (IEEE Cat. No. 04CH37566)*, vol. 3. (IEEE, 2004), pp. 2149–2154

63. E. Todorov, T. Erez, Y. Tassa, Mujoco: a physics engine for model-based control, in *IEEE/RSJ International Conference on Intelligent Robots and Systems* (IEEE, 2012), pp. 5026–5033

64. E. Rohmer, S.P. Singh, M. Freese, V-REP: a versatile and scalable robot simulation framework, in *IEEE/RSJ International Conference on Intelligent Robots and Systems* (IEEE, 2013), pp. 1321–1326

65. A. Dosovitskiy, G. Ros, F. Codevilla, A. Lopez, V. Koltun, Carla: an open urban driving simulator, in *Conference on Robot Learning* (PMLR, 2017), pp. 1–16

66. S. Shah, D. Dey, C. Lovett, A. Kapoor, AirSim: high-fidelity visual and physical simulation for autonomous vehicles, in *Field and Service Robotics: Results of the 11th International Conference* (Springer, Berlin, 2018), pp. 621–635

67. O. Michel, Cyberbotics ltd. webots™: professional mobile robot simulation. Int. J. Adv. Robot. Syst. **1**(1), 5 (2004)

68. S. James, M. Freese, A.J. Davison, PyRep: bringing V-REP to deep robot learning (2019). arXiv:1906.11176

69. M. Kirtas, K. Tsampazis, N. Passalis, A. Tefas, Deepbots: a webots-based deep reinforcement learning framework for robotics, in *Artificial Intelligence Applications and Innovations: 16th IFIP WG 12.5 International Conference, AIAI, Neos Marmaras, Greece, June 5–7, 2020, Proceedings, Part II 16* (Springer, Berlin, 2020), pp. 64–75

70. B. Dariush, M. Gienger, A. Arumbakkam, Y. Zhu, B. Jian, K. Fujimura, C. Goerick, Online transfer of human motion to humanoids. Int. J. Humanoid Rob. **6**(02), 265–289 (2009)

71. A. Bajcsy, D.P. Losey, M.K. O'Malley, A.D. Dragan, Learning from physical human corrections, one feature at a time, in *Proceedings of the 2018 ACM/IEEE International Conference on Human-Robot Interaction* (2018), pp. 141–149

72. D. Lee, C. Ott, Incremental kinesthetic teaching of motion primitives using the motion refinement tube. Auton. Robot. **31**, 115–131 (2011)
73. S. Schaal, Dynamic movement primitives-a framework for motor control in humans and humanoid robotics, in *Adaptive Motion of Animals and Machines* (Springer, Berlin, 2006), pp. 261–280
74. J. Campbell, H.B. Amor, Bayesian interaction primitives: a SLAM approach to human-robot interaction, in *Conference on Robot Learning* (PMLR, 2017), pp. 379–387
75. T. Welschehold, C. Dornhege, W. Burgard, Learning mobile manipulation actions from human demonstrations, in *2017 IEEE/RSJ International Conference on Intelligent Robots and Systems (IROS)* (IEEE, 2017), pp. 3196–3201
76. T. Welschehold, N. Abdo, C. Dornhege, W. Burgard, Combined task and action learning from human demonstrations for mobile manipulation applications, in *2019 IEEE/RSJ International Conference on Intelligent Robots and Systems (IROS)* (IEEE, 2019), pp. 4317–4324
77. B. Akgun, M. Cakmak, J.W. Yoo, A.L. Thomaz, Trajectories and keyframes for kinesthetic teaching: a human-robot interaction perspective, in *Proceedings of the Seventh Annual ACM/IEEE International Conference on Human-Robot Interaction* (2012), pp. 391–398
78. C. Pérez-D'Arpino, J.A. Shah, C-learn: learning geometric constraints from demonstrations for multi-step manipulation in shared autonomy, in *2017 IEEE International Conference on Robotics and Automation (ICRA)* (IEEE, 2017), pp. 4058–4065
79. T. Yamamoto, K. Terada, A. Ochiai, F. Saito, Y. Asahara, K. Murase, Development of human support robot as the research platform of a domestic mobile manipulator. ROBOMECH J **6**(1), 1–15 (2019)
80. C. Lynch, M. Khansari, T. Xiao, V. Kumar, J. Tompson, S. Levine, P. Sermanet, Learning latent plans from play, in *Conference on Robot Learning* (PMLR, 2020), pp. 1113–1132
81. K. Yamane, Y. Ariki, J. Hodgins, Animating non-humanoid characters with human motion data, in *Proceedings of the 2010 ACM SIGGRAPH/Eurographics Symposium on Computer Animation* (2010), pp. 169–178
82. G. Tevet, S. Raab, B. Gordon, Y. Shafir, D. Cohen-Or, A.H. Bermano, Human motion diffusion model (2022). arXiv:2209.14916
83. M. Zhang, Z. Cai, L. Pan, F. Hong, X. Guo, L. Yang, Z. Liu, Motiondiffuse: text-driven human motion generation with diffusion model. IEEE Trans. Pattern Anal. Mach. Intell. **46**(6), 4115–4128 (2024)
84. X. Chen, B. Jiang, W. Liu, Z. Huang, B. Fu, T. Chen, and G. Yu, Executing your commands via motion diffusion in latent space, in *Proceedings of the IEEE/CVF Conference on Computer Vision and Pattern Recognition* (2023), pp. 18 000–18 010
85. Y. Yuan, J. Song, U. Iqbal, A. Vahdat, J. Kautz, Physdiff: physics-guided human motion diffusion model, in *Proceedings of the IEEE/CVF International Conference on Computer Vision* (2023), pp. 16 010–16 021
86. T.Z. Zhao, V. Kumar, S. Levine, C. Finn, Learning fine-grained bimanual manipulation with low-cost hardware (2023). arXiv:2304.13705
87. K. Black, N. Brown, D. Driess, A. Esmail, M. Equi, C. Finn, N. Fusai, L. Groom, K. Hausman, B. Ichter, et al., $\pi_0$: a vision-language-action flow model for general robot control (2024). arXiv:2410.24164
88. O.M. Team, D. Ghosh, H. Walke, K. Pertsch, K. Black, O. Mees, S. Dasari, J. Hejna, T. Kreiman, C. Xu, et al., Octo: an open-source generalist robot policy (2024). arXiv:2405.12213
89. M.J. Kim, K. Pertsch, S. Karamcheti, T. Xiao, A. Balakrishna, S. Nair, R. Rafailov, E. Foster, G. Lam, P. Sanketi, et al., Openvla: an open-source vision-language-action model (2024). arXiv:2406.09246
90. M.R. Cutkosky et al., On grasp choice, grasp models, and the design of hands for manufacturing tasks. IEEE Trans. Robot. Autom. **5**(3), 269–279 (1989)

91. S.B. Kang, K. Ikeuchi, Toward automatic robot instruction from perception-mapping human grasps to manipulator grasps. IEEE Trans. Robot. Autom. **13**(1), 81–95 (1997)

92. T. Feix, J. Romero, H.-B. Schmiedmayer, A.M. Dollar, D. Kragic, The grasp taxonomy of human grasp types. IEEE Trans. Hum.-Mach. Syst. **46**(1), 66–77 (2015)

93. K. Ikeuchi, T. Kanade, Modelling sensors: toward automatic generation of object recognition program. Comput. Vis., Graph., Image Process. **48**(1), 50–79 (1989)

94. N. Wake, K. Sasabuchi, K. Ikeuchi, Grasp-type recognition leveraging object affordance, in *HOBI Workshop, IEEE International Symposium on Robot and Human Interactive Communication (RO-MAN)* (2020)

95. R.C. Brost, Automatic grasp planning in the presence of uncertainty. Int. J. Robot. Res. **7**(1), 3–17 (1988)

96. A. Bicchi, V. Kumar, Robotic grasping and contact: a review, in *Proceedings, ICRA. Millennium Conference. IEEE International Conference on Robotics and Automation. Symposia Proceedings (Cat. No. 00CH37065)*, vol. 1 (IEEE 2000), pp. 348–353

97. B.K. Horn, K. Ikeuchi, The mechanical manipulation of randomly oriented parts. Sci. Am. **251**(2), 100–113 (1984)

98. C. Ferrari, J. Canny, et al., Planning optimal grasps, in *Proceedings, 1992 IEEE International Conference on Robotics and Automation*, vol. 3 (IEEE, 1992), pp. 2290–2295

99. K. Ikeuchi, T. Kanade, Automatic generation of object recognition programs. Proc. IEEE **76**(8), 1016–1035 (1988)

100. A.T. Miller, S. Knoop, H.I. Christensen, P.K. Allen, Automatic grasp planning using shape primitives, in *2003 IEEE International Conference on Robotics and Automation (Cat. No. 03CH37422)*, vol. 2 (IEEE, 2003), pp. 1824–1829

101. C. Goldfeder, P.K. Allen, C. Lackner, R. Pelossof, Grasp planning via decomposition trees, in *Proceedings 2007 IEEE International Conference on Robotics and Automation* (IEEE, 2007), pp. 4679–4684

102. K. Huebner, S. Ruthotto, D. Kragic, Minimum volume bounding box decomposition for shape approximation in robot grasping, in *2008 IEEE International Conference on Robotics and Automation* (IEEE, 2008), pp. 1628–1633

103. M. Dogar, K. Hsiao, M. Ciocarlie, S. Srinivasa, Physics-based grasp planning through clutter. *Robotics: Science and Systems VIII* (2012), p. 57

104. A.T. Miller, P.K. Allen, Graspit! a versatile simulator for robotic grasping. IEEE Robot. Autom. Mag. **11**(4), 110–122 (2004)

105. I. Lenz, H. Lee, A. Saxena, Deep learning for detecting robotic grasps. Int. J. Robot. Res. **34**(4–5), 705–724 (2015)

106. J. Redmon, A. Angelova, Real-time grasp detection using convolutional neural networks, in *IEEE International Conference on Robotics and Automation (ICRA)* (IEEE, 2015), pp. 1316–1322

107. Z. Wang, Z. Li, B. Wang, H. Liu, Robot grasp detection using multimodal deep convolutional neural networks. Adv. Mech. Eng. **8**(9), 1687814016668077 (2016)

108. S. Levine, P. Pastor, A. Krizhevsky, J. Ibarz, D. Quillen, Learning hand-eye coordination for robotic grasping with deep learning and large-scale data collection. Int. J. Robot. Res. **37**(4–5), 421–436 (2018)

109. D. Kalashnikov, A. Irpan, P. Pastor, J. Ibarz, A. Herzog, E. Jang, D. Quillen, E. Holly, M. Kalakrishnan, V. Vanhoucke, et al., Scalable deep reinforcement learning for vision-based robotic manipulation, in *Conference on Robot Learning* (PMLR, 2018), pp. 651–673

110. K. Rao, C. Harris, A. Irpan, S. Levine, J. Ibarz, M. Khansari, Rl-cyclegan: reinforcement learning aware simulation-to-real, in *Proceedings of the IEEE/CVF Conference on Computer Vision and Pattern Recognition* (2020), pp. 11 157–11 166

111. R. Calandra, A. Owens, D. Jayaraman, J. Lin, W. Yuan, J. Malik, E.H. Adelson, S. Levine, More than a feeling: learning to grasp and regrasp using vision and touch. IEEE Robot. Autom. Lett. **3**(4), 3300–3307 (2018)
112. F.R. Hogan, M. Bauza, O. Canal, E. Donlon, A. Rodriguez, Tactile regrasp: grasp adjustments via simulated tactile transformations, in *2018 IEEE/RSJ International Conference on Intelligent Robots and Systems (IROS)* (IEEE, 2018), pp. 2963–2970
113. N. Kamakura, *Shape of Hand and Movement of Hand* (Ishiyaku Publisher, Tokyo, 1989) (in Japanese)
114. S.B. Kang, K. Ikeuchi, Toward automatic robot instruction from perception-recognizing a grasp from observation. IEEE Trans. Robot. Autom. **9**(4), 432–443 (1993)
115. I.M. Bullock, R.R. Ma, A.M. Dollar, A hand-centric classification of human and robot dexterous manipulation. IEEE Trans. Haptics **6**(2), 129–144 (2012)
116. M. Vergara, J.L. Sancho-Bru, V. Gracia-Ibáñez, A. Pérez-González, An introductory study of common grasps used by adults during performance of activities of daily living. J. Hand Ther. **27**(3), 225–234 (2014)
117. M.A. Arbib, Coordinated control programs for movement of the hand, in *Hand Function and the Neocortex* (1985), pp. 111–129
118. H. Li, Y. Zhang, Y. Li, H. He, Learning task-oriented dexterous grasping from human knowledge, in *2021 IEEE International Conference on Robotics and Automation (ICRA)* (IEEE, 2021), pp. 6192–6198
119. H. Li, D. Tran, X. Zhang, H. He, Knowledge augmentation and task planning in large language models for dexterous grasping, in *2023 IEEE-RAS 22nd International Conference on Humanoid Robots (Humanoids)* (IEEE, 2023), pp. 1–8
120. P. Mandikal, K. Grauman, Learning dexterous grasping with object-centric visual affordances, in *IEEE International Conference on Robotics and Automation (ICRA)* (IEEE, 2021), pp. 6169–6176
121. Z. Qin, K. Fang, Y. Zhu, L. Fei-Fei, S. Savarese, Keto: learning keypoint representations for tool manipulation, in *2020 IEEE International Conference on Robotics and Automation (ICRA)* (IEEE, 2020), pp. 7278–7285
122. M. Kokic, D. Kragic, J. Bohg, Learning task-oriented grasping from human activity datasets. IEEE Robot. Autom. Lett. **5**(2), 3352–3359 (2020)
123. S. Brahmbhatt, C. Ham, C.C. Kemp, J. Hays, ContactDB: analyzing and predicting grasp contact via thermal imaging, in *Proceedings of the IEEE/CVF Conference on Computer Vision and Pattern Recognition* (2019), pp. 8709–8719
124. S. Brahmbhatt, C. Tang, C.D. Twigg, C.C. Kemp, J. Hays, Contactpose: a dataset of grasps with object contact and hand pose, in *Computer Vision-ECCV, 16th European Conference, Glasgow, UK, August 23–28, 2020, Proceedings, Part XIII 16* (Springer, Berlin, 2020), pp. 361–378
125. P. Mandikal, K. Grauman, Dexvip: learning dexterous grasping with human hand pose priors from video, in *Conference on Robot Learning* (PMLR, 2022), pp. 651–661
126. H. Merzić, M. Bogdanović, D. Kappler, L. Righetti, J. Bohg, Leveraging contact forces for learning to grasp, in *International Conference on Robotics and Automation (ICRA)* (IEEE, 2019), pp. 3615–3621
127. B. Wu, I. Akinola, J. Varley, P. Allen, MAT: multi-fingered adaptive tactile grasping via deep reinforcement learning (2019). arXiv:1909.04787
128. A. Koenig, Z. Liu, L. Janson, R.D. Howe, Tactile grasp refinement using deep reinforcement learning and analytic grasp stability metrics. *CoRR* (2021)
129. A. Rajeswaran, V. Kumar, A. Gupta, G. Vezzani, J. Schulman, E. Todorov, S. Levine, Learning complex dexterous manipulation with deep reinforcement learning and demonstrations (2017). arXiv:1709.10087

130. E. Valarezo Anazco, P. Rivera Lopez, N. Park, J. Oh, G. Ryu, M.A. Al-Antari, T.-S. Kim, Natural object manipulation using anthropomorphic robotic hand through deep reinforcement learning and deep grasping probability network. Appl. Intell. **51**, 1041–1055 (2021)

131. K. Fang, Y. Zhu, A. Garg, A. Kurenkov, V. Mehta, L. Fei-Fei, S. Savarese, Learning task-oriented grasping for tool manipulation from simulated self-supervision. Int. J. Robot. Res. **39**(2–3), 202–216 (2020)

132. P. Vinayavekhin, S. Kudoh, K. Ikeuchi, Towards an automatic robot regrasping movement based on human demonstration using tangle topology, in *2011 IEEE International Conference on Robotics and Automation* (IEEE, 2011), pp. 3332–3339

133. H.T. Kuhn, W.L. Inequalities, *Related Systems, Annals of Mathematic Studies* (Princeton University Press, EEUU, 1956)

134. M. Ohwovoriole, B. Roth, An extension of screw theory. J. Mech. Des. **103**(4), 725–735 (1981)

135. K. Ikeuchi, N. Wake, K. Sasabuchi, J. Takamatsu, Semantic constraints to represent common sense required in household actions for multimodal learning-from-observation robot. Int. J. Robot. Res. **43**(2), 134–170 (2024)

136. R.P. Paul, *Robot Manipulators: Mathematics, Programming, and Control: The Computer Control of Robot Manipulators* (Richard Paul, 1981)

137. N. Hogan, Impedance control: an approach to manipulation, in *American Control Conference* (IEEE, 1984), pp. 304–313

138. M.T. Mason, Compliance and force control for computer controlled manipulators. IEEE Trans. Syst. Man Cybern. **11**(6), 418–432 (1981)

139. A. Yahya, A. Li, M. Kalakrishnan, Y. Chebotar, S. Levine, Collective robot reinforcement learning with distributed asynchronous guided policy search, in *2017 IEEE/RSJ International Conference on Intelligent Robots and Systems (IROS)* (IEEE, 2017), pp. 79–86

140. S. Levine, C. Finn, T. Darrell, P. Abbeel, End-to-end training of deep visuomotor policies. J. Mach. Learn. Res. **17**(39), 1–40 (2016)

141. Y. Urakami, A. Hodgkinson, C. Carlin, R. Leu, L. Rigazio, P. Abbeel, DoorGym: a scalable door opening environment and baseline agent (2019). arXiv:1908.01887

142. S. Gu, E. Holly, T. Lillicrap, S. Levine, Deep reinforcement learning for robotic manipulation with asynchronous off-policy updates, in *IEEE International Conference on Robotics and Automation (ICRA)* (IEEE, 2017), pp. 3389–3396

143. A.V. Nair, V. Pong, M. Dalal, S. Bahl, S. Lin, S. Levine, Visual reinforcement learning with imagined goals, in *Advances in Neural Information Processing Systems*, vol. 31 (2018)

144. Y. Sun, L. Zhang, O. Ma, Force-vision sensor fusion improves learning-based approach for self-closing door pulling. IEEE Access **9**, 137 188–137 197 (2021)

145. N. Vithayathil Varghese, Q.H. Mahmoud, A survey of multi-task deep reinforcement learning. Electronics **9**(9), 1363 (2020)

146. Y. Duan, M. Andrychowicz, B. Stadie, O. Jonathan Ho, J. Schneider, I. Sutskever, P. Abbeel, W. Zaremba, One-shot imitation learning, in *Advances in Neural Information Processing Systems*, vol. 30 (2017)

147. T. Zhang, Z. McCarthy, O. Jow, D. Lee, X. Chen, K. Goldberg, P. Abbeel, Deep imitation learning for complex manipulation tasks from virtual reality teleoperation, in *IEEE International Conference on Robotics and Automation (ICRA)* (IEEE, 2018), pp. 5628–5635

148. E. Johns, Coarse-to-fine imitation learning: robot manipulation from a single demonstration, in *IEEE International Conference on Robotics and Automation (ICRA)* (IEEE, 2021), pp. 4613–4619

149. N.J. Cho, S.H. Lee, J.B. Kim, I.H. Suh, Learning, improving, and generalizing motor skills for the peg-in-hole tasks based on imitation learning and self-learning. Appl. Sci. **10**(8), 2719 (2020)

150. S.A. Mehta, R.S. Zarrin, On the feasibility of a mixed-method approach for solving long horizon task-oriented dexterous manipulation, in *2024 IEEE-RAS 23rd International Conference on Humanoid Robots (Humanoids)* (IEEE, 2024), pp. 949–956

151. T. Buamanee, M. Kobayashi, Y. Uranishi, H. Takemura, Bi-act: bilateral control-based imitation learning via action chunking with transformer, in *2024 IEEE International Conference on Advanced Intelligent Mechatronics (AIM)* (IEEE, 2024), pp. 410–415

152. A. Lee, I. Chuang, L.-Y. Chen, I. Soltani, Interact: inter-dependency aware action chunking with hierarchical attention transformers for bimanual manipulation (2024). arXiv:2409.07914

153. A. O'Neill, A. Rehman, A. Maddukuri, A. Gupta, A. Padalkar, A. Lee, A. Pooley, A. Gupta, A. Mandlekar, A. Jain, et al., Open X-embodiment: robotic learning datasets and RT-X models: open X-embodiment collaboration[0], in *2024 IEEE International Conference on Robotics and Automation (ICRA)* (IEEE, 2024), pp. 6892–6903

154. A. Brohan, N. Brown, J. Carbajal, Y. Chebotar, X. Chen, K. Choromanski, T. Ding, D. Driess, A. Dubey, C. Finn, et al., RT-2: vision-language-action models transfer web knowledge to robotic control (2023). arXiv:2307.15818

155. Z. Wang, Z. Zhou, J. Song, Y. Huang, Z. Shu, L. Ma, Towards testing and evaluating vision-language-action models for robotic manipulation: an empirical study (2024). arXiv:2409.12894

156. C. Aeronautiques, A. Howe, C. Knoblock, I.D. McDermott, A. Ram, M. Veloso, D. Weld, D.W. Sri, A. Barrett, D. Christianson, et al., *PDDL| The Planning Domain Definition Language* (Technical Report, 1998)

157. B. Bonet, H. Geffner, Planning as heuristic search. Artif. Intell. **129**(1–2), 5–33 (2001)

158. M. Helmert, The fast downward planning system. J. Artif. Intell. Res. **26**, 191–246 (2006)

159. A. Gudimella, R. Story, M. Shaker, R. Kong, M. Brown, V. Shnayder, M. Campos, Deep reinforcement learning for dexterous manipulation with concept networks (2017). arXiv:1709.06977

160. M. Dalal, D. Pathak, R.R. Salakhutdinov, Accelerating robotic reinforcement learning via parameterized action primitives, in *Advances in Neural Information Processing Systems*, vol. 34 (2021), pp. 21 847–21 859

161. S. Nasiriany, H. Liu, Y. Zhu, Augmenting reinforcement learning with behavior primitives for diverse manipulation tasks, in *2022 International Conference on Robotics and Automation (ICRA)* (IEEE, 2022), pp. 7477–7484

162. A. Hiranaka, M. Hwang, S. Lee, C. Wang, L. Fei-Fei, J. Wu, R. Zhang, Primitive skill-based robot learning from human evaluative feedback, in *2023 IEEE/RSJ International Conference on Intelligent Robots and Systems (IROS)* (IEEE, 2023), pp. 7817–7824

163. U.A. Mishra, S. Xue, Y. Chen, D. Xu, Generative skill chaining: Long-horizon skill planning with diffusion models, in *Conference on Robot Learning* (PMLR, 2023), pp. 2905–2925

164. K. Shirai, C.C. Beltran-Hernandez, M. Hamaya, A. Hashimoto, S. Tanaka, K. Kawaharazuka, K. Tanaka, Y. Ushiku, S. Mori, Vision-language interpreter for robot task planning, in *2024 IEEE International Conference on Robotics and Automation (ICRA)* (IEEE, 2024), pp. 2051–2058

165. A. Chowdhery, S. Narang, J. Devlin, M. Bosma, G. Mishra, A. Roberts, P. Barham, H.W. Chung, C. Sutton, S. Gehrmann et al., Palm: scaling language modeling with pathways. J. Mach. Learn. Res. **24**(240), 1–113 (2023)

166. T. Yu, D. Quillen, Z. He, R. Julian, K. Hausman, C. Finn, S. Levine, Meta-world: a benchmark and evaluation for multi-task and meta reinforcement learning, in *Conference on Robot Learning* (PMLR, 2020), pp. 1094–1100

167. A. Gupta, V. Kumar, C. Lynch, S. Levine, K. Hausman, Relay policy learning: solving long-horizon tasks via imitation and reinforcement learning (2019). arXiv:1910.11956

168. Y. Zhu, J. Wong, A. Mandlekar, R. Martín-Martín, A. Joshi, S. Nasiriany, Y. Zhu, Robosuite: a modular simulation framework and benchmark for robot learning (2020). arXiv:2009.12293

169. D. Marr, *Vision: A Computational Investigation into the Human Representation and Processing of Visual Information* (MIT Press, 2010)
170. L.G. Roberts, Machine perception of three-dimensional solids. Ph.D. dissertation, Massachusetts Institute of Technology (1963)
171. W.E.L. Grimson, *Object Recognition by Computer: The Role of Geometric Constraints* (MIT Press, 1991)
172. R.A. Brooks, Symbolic reasoning among 3-d models and 2-d images. Artif. Intell. **17**(1–3), 285–348 (1981)
173. D.G. Lowe, Object recognition from local scale-invariant features, in *Proceedings of the Seventh IEEE International Conference on Computer Vision*, vol. 2 (IEEE, 1999), pp. 1150–1157
174. J. Redmon, A. Farhadi, Yolo9000: better, faster, stronger, in *Proceedings of the IEEE Conference on Computer Vision and Pattern Recognition* (2017), pp. 7263–7271
175. J. Deng, W. Dong, R. Socher, L.-J. Li, K. Li, L. Fei-Fei, Imagenet: a large-scale hierarchical image database, in *IEEE Conference on Computer Vision and Pattern Recognition* (IEEE, 2009), pp. 248–255
176. S. Ren, K. He, R. Girshick, J. Sun, Faster R-CNN: towards real-time object detection with region proposal networks. IEEE Trans. Pattern Anal. Mach. Intell. **39**(6), 1137–1149 (2016)
177. K. Ikeuchi, M. Kawade, T. Suehiro, Assembly task recognition with planar, curved and mechanical contacts, in *1993 Proceedings IEEE International Conference on Robotics and Automation* (IEEE, 1993), pp. 688–694
178. J. Miura, K. Ikeuchi, Task-oriented generation of visual sensing strategies in assembly tasks. IEEE Trans. Pattern Anal. Mach. Intell. **20**(2), 126–138 (1998)
179. C.C. Adams, *The Knot Book: An Elementary Introduction to the Mathematical Theory of Knots* (American Mathematical Society, 2004)
180. M. Ochiaki, Y.S.E. Toyoda, *Computer Aided Knot Thoery* (Makino Shoten, 1996). In Japanese
181. A.P. Witkin, Scale-space filtering, in *Readings in Computer Vision* 9Elsevier, 1987), pp. 329–332
182. T. Shiratori, K. Ikeuchi, Synthesis of dance performance based on analyses of human motion and music. Inf. Media Technol. **3**(4), 834–847 (2008)
183. T. Okamoto, T. Shiratori, S. Kudoh, S. Nakaoka, K. Ikeuchi, Toward a dancing robot with listening capability: Keypose-based integration of lower-, middle-, and upper-body motions for varying music tempos. IEEE Trans. Rob. **30**(3), 771–778 (2014)
184. N.S. Pollard, J.K. Hodgins, M.J. Riley, C.G. Atkeson, Adapting human motion for the control of a humanoid robot, in *Proceedings, IEEE International Conference on Robotics and Automation (Cat. No. 02CH37292)*, vol. 2 (IEEE, 2002), pp. 1390–1397
185. M. Vukobratovic, D. Juricic, Contribution to the synthesis of biped gait. IEEE Trans. Biomed. Eng. **1**, 1–6 (1969)
186. S. Kajita, F. Kanehiro, K. Kaneko, K. Fujiwara, K. Harada, K. Yokoi, H. Hirukawa, Biped walking pattern generation by using preview control of zero-moment point, in *IEEE International Conference on Robotics and Automation (Cat. No. 03CH37422)*, vol. 2 (IEEE, 2003), pp. 1620–1626
187. M. Nishiwaki, S. Kagami, M. Inaba, H. Inoue, High-speed generation of dynamically stable trajectories for humanoids through linear decoupling and discretization of ZMP derivation, in *The 18th Annual Conference of the Robotics Society of Japan* (RSJ, 2000). In Japanese
188. Y. Tamiya, M. Inaba, H. Inoue, Real-time dynamic balance compensation using the entire body in single-leg standing actions of humanoid robots. J. Robot. Soc. Jpn. **17**(2), 268–274 (1999). (in Japanese)

189. T. Isozumi, K. Akachi, M. Hirata, K. Kaneko, S. Kajita, H. Hirukawa, Development of the humanoid robot HRP-2. J. Robot. Soc. Jpn. **22**(8), 1004–1012 (2004). (in Japanese)
190. N. Wake, A. Kanehira, J. Takamatsu, K. Sasabuchi, K. Ikeuchi, VLM-driven behavior tree for context-aware task planning (2025). arXiv:2501.03968
191. R. Reddy, To dream the possible dream, in *ACM Turing Award Lectures* (2007), p. 1994

# Index

FSC
www.fsc.org
MIX
Papier | Fördert
gute Waldnutzung
FSC® C083411

Zeitfracht Medien GmbH
Ferdinand-Jühlke-Straße 7
99095 Erfurt, Deutschland
produktsicherheit@kolibri360.de